AQUATIC FOOD QUALITY AND SAFETY ASSESSMENT METHODS

The textbook on "Aquatic Food Quality and Safety Assessment Methods" explains on the methods and procedures adopted for testing the quality and safety of aquatic food products. The analytical techniques available for testing the chemical constituents of aquatic food with separate chapters on the analysis of lipids, proteins, vitamins, and minerals are exhaustively given to determine their nutritional quality. The various methods for sensory, physical, biochemical and microbiological quality assessments of aquatic food are explicitly given with detailed protocols for easy adoption. Special chapters covering the chemical contaminants and permitted additives for residue monitoring are dealt, as they are important food safety requirements. This book will be very helpful for the food quality control technologists, food analysts, research scholars, and fisheries professionals as a holistic guide on a variety of testing procedures for facile adoption to meet the food safety and quality regulatory requirements.

Prof. R. Jeya Shakila, Head of the Department of Fish Quality Assurance and Management, Fisheries College & Research Institute, Tamil Nadu Dr. J. Jayalalithaa Fisheries University has been working in the fields of fish biochemistry, quality, safety and authenticity for more than 21 years. She worked in Defence Food Research Laboratory (DFRL), Mysore for her Ph.D. degree in the area of Biogenic amines from fish and fishery products.

Prof. G. Jeyasekaran, Director of Research of Tamil Nadu Dr. J. Jayalalithaa Fisheries University, has been working for more than 33 years and involved in teaching, and research activities pertaining to fish quality, safety, authenticity, microbiology and biotechnology. Being the first Director of Research of the University, he was involved in the overall development of the research programmes.

AQUATIC FOOD QUALITY AND SAFETY ASSESSMENT METHODS

Prof. R. JEYA SHAKILA

and

Prof. G. JEYASEKARAN

Department of Fish Quality Assurance and Management
Tamil Nadu Dr. J. Jayalalithaa Fisheries University
Thoothukudi 628 008, Tamil Nadu, INDIA

CRC Press
Taylor & Francis Group
Boca Raton London New York

CRC Press is an imprint of the
Taylor & Francis Group, an **informa** business

NARENDRA PUBLISHING HOUSE
DELHI (INDIA)

First published 2021
by CRC Press
2 Park Square, Milton Park, Abingdon, Oxon, OX14 4RN

and by CRC Press
6000 Broken Sound Parkway NW, Suite 300, Boca Raton, FL 33487-2742

© 2021 Narendra Publishing House

CRC Press is an imprint of Informa UK Limited

Print edition not for sale in South Asia (India, Sri Lanka, Nepal, Bangladesh, Pakistan or Bhutan).

British Library Cataloguing-in-Publication Data
A catalogue record for this book is available from the British Library

Library of Congress Cataloging-in-Publication Data
A catalog record for this book has been requested

ISBN: 978-0-367-61946-6 (hbk)
ISBN: 978-1-003-10719-4 (ebk)

**NARENDRA PUBLISHING
HOUSE
DELHI (INDIA)**

Contents

Preface

Twenty years ago, when we started working together in a small laboratory on few experiments to assess the safety and quality of fish, we never realized that it will emerge into a new kind of science "Fish Quality Assurance and Management". More and more, we investigated, we realized that the theme on fish quality and safety is one of the emerging areas of science, the world is currently moving ahead. We have seen the beginning of a partition in the existing subject "Fish Processing Technology", when we got a clue that fish quality, safety, authenticity, and traceability is towards a whole new kind of science.

We are glad to bring this book on "Aquatic Food Quality and Safety Assessment Methods" for the scientific community focussed on this area of science. This book is the culmination of our twenty years of experience working together. We never expected that it would take such a long time to consolidate our works altogether to prepare this book vastly for the community. We have added some additional information on each topic to make it readable in a cohesive manner.

In our early career path, as Professors cum Scientists we have published several research papers out of our research projects in scientific literatures. But, although our publications wrote in scientific journals are very well received as assessed by our citations, we gradually realized that the methodologies scattered across the journals on all aspects of sensory, physical, chemical and microbiological analysis; and the residue testing must be consolidated in the form a book. It provides a better handbook for the scholars majoring in this new area of science, which we begin to build further.

We have worked out quietly to bring this book taking into account all our practical knowledge for the research scholars, food scientists, food analysts and food quality control technologists, who need a consolidated book consisting of all standard protocols to analyze the aquatic food for its nutritional aspects, quality and safety requirements. We are sure that this book provides all the relevant standard methodologies adopted for testing the chemical constituents, biochemical indices, pathogenic bacteria detection, and chemical residue monitoring as per the regulatory requirements of USFDA and EU.

We wish to thank few scholars and students, who have helped us to compile our scientific literatures into this book with appropriate references. We wish the readers to make use of this book in actual practice in the laboratory for achieving good results.

Authors

CHAPTER 1

INTRODUCTION

Aquatic food is a common term used to refer to the food of aquatic origin, which includes fish, shellfish, and seaweeds. The global fish production recorded ias 170.9 million tons in 2018, with marine capture fisheries contributing to 90.9 million tons and aquaculture to 80.0 million tons. The expansion in inland and brackishwater aquaculture, as well as mariculture practices over the years, has contributed to the significant growth of the total fish production. As the fish inhabits different niche, it is more appropriate to refer the foods of aquatic origin as 'aquatic foods'. This term conveys a better understanding of all the food organisms taken from marine, freshwater, and brackishwater origin. The major foods of aquatic origin include fish, shellfish (including crustaceans and mollusks), and algae (including seaweeds).

Fish belongs to the kingdom Animalia and the phylum Chordata, and further classified into three classes, namely Agnatha (jawless fish), Chondrichthyes (cartilaginous fish) and Osteichthyes (bony fish). Most of the food fish mainly belong to the latter two classes. Fish of the class, Chondrichthyes, have a skeleton made of cartilage. Sharks, rays, and skates belong to this class and referred to as elasmobranchs. Fish of the class, Osteichthyes, have a skeleton composed of bone. They are further classified as a ray-finned group (perch and catfish) and a lobe-finned group (lungfish). Crustaceans belong to the phylum Arthropoda and sub-phylum Crustacea. They are invertebrates that have a segmented body covered with chitinous exoskeleton. Shrimps, crabs, lobsters, and crayfish are important among this group. Mollusks belong to the phylum Mollusca, and they are soft-bodied animals that have an internal or external protective shell. Bivalvia and Cephalopoda are two major classes of invertebrates consumed as food. Oysters, clams, cockles, mussels, and scallops are important from a commercial point of view under the class, Bivalvia. Squids, octopus, and cuttlefish are the main groups consumed as food under the class, Cephalopoda. Seaweeds are not plants, instead marine algae belonging to the kingdom, Protista. There are three distinct groups, namely brown algae (Phaeophyta), green algae (Chlorophyta) and red algae (Rhodophyta) that are edible. Seaweeds are being traditionally used as food by the people of China, Japan, and Korea.

Wild-capture or culture methods are employed to take aquatic foods from nature. Wild fish are caught mainly by gill nets, long lines, trawl nets, and purse seines. Cultured fish are harvested by cast nets or by complete drain operation. Whatever be the method of harvest, aquatic foods rapidly undergo spoilage. The reasons behind rapid deterioration are a low concentration of connective tissue proteins and large quantities of nitrogen-containing extractives in fish. The spoilage occurs immediately after harvest, and the onset of rigor occurs within 6-10 h, if fish are held at ambient temperature. The widely practiced method onboard and onshore to delay the spoilage is the preservation of fish in ice. The iced fish is then subjected to processing by freezing or by thermal processing or by value addition. Fish can be processed in raw, frozen, cooked, smoked, dried, marinated, pickled, fermented, or heat-processed ready-to-eat (RTE) forms, depending on the consumer choice.

Fish processors or fish business operators (FBOs) need to supply safe and quality aquatic food to the consumers. Mostly fish are placed in the market either directly by fishers or by FBOs. In developing countries, fishing communities involve themselves directly in marketing, while in well-developed countries, FBOs perform the job. The global seafood market is currently valued at 125.2 billion dollars in 2017 and expected to grow considerably to 155.3 billion dollars in 2023.

The huge market value for the aquatic foods and the issues related to their rapid decomposition made the regulatory bodies to introduce hygiene requirements at the production and marketing stages. The European Council (EC) laid down regulations for the marketing of fishery products way back in 1981 by fixing standards (EEC No. 3796/81). Rules are made for certain fishery products to place them on the market to promote intra-community trade by the Council Directive (91/493/EEC). Common marketing standards are laid down for certain fishery products such as saltwater fish and crustaceans by categorizing them based on freshness and size (EC No. 2406/96). Specific hygiene rules for foodstuff are laid down in the European Parliament by giving requirements for handling of fishery products, frozen, mechanically separated and processing fishery products (EC No. 853/2004). The health standards generally comprise of organoleptic, histamine, TVB-N, parasites, and marine bio-toxin checks. Rules are made for packaging, storage, and transportation of fishery products also by the EC.

Implementation of stringent rules, regulations, and standards for fishery products by the EU legislation led other nations also to formulate their legislations to implement health standards for fishery products. Codex Alimentarius Commission provides an international standard for fishery products for adoption by third world countries. Codex Alimentarius also provides several codes of practices and standards (www.codexalimentarius.org). The US Food and Drug Administration (FDA) also operates a mandatory safety programme for fish and fishery products. FDA publishes Fish and Fisheries Products Hazards and Controls Guidance by compiling up-to-date science and policy on the hazards and effective control measures to prevent their occurrence. The fourth edition of this document is available currently as part of the regulatory programme around the World (www.fda.org). This document provides information on hazard analysis, HACCP plan development, potential hazards, pathogens, natural toxins, decomposition related hazards, environmental chemical contaminants, pesticides, aquaculture drugs, heavy metals, food allergens, and physical hazards. FDA provides regulations concerning fish and fishery products (21CFR Part 123), molluscan shellfish (21CFR Part 1240.60), and crabmeat (21 CFR Part 102). There are specific guidance documents for controlling the hazard of *Clostridium botulinum* growth and toxin production in reduced oxygen packaged fish and fishery products, including refrigerated, vacuum-packed crawfish tail meat; control of *Listeria monocytogenes* in RTE foods, fraudulent practice of including glaze as part of the weight of frozen foods, control of pathogens in cold-smoked fish and so on. FDA also specifies additional information about seafood species substitution and aquacultured food.

In India, Food Safety and Standards Authority of India (FSSAI) issued vertical standards for fish and fishery products through Food Safety and Standards (Food Products Standards and Food Additives) Regulation for frozen shrimp, frozen cephalopods, smoked fishery products, ready-to-eat fish curry in retortable pouches, sardine oil, edible fish powder, fish pickles, frozen minced fish meat, freeze-dried prawns and frozen clam meat. The FSSAI has brought down microbiological standards for fish and fishery products through a gazette notification of Food Safety and Standards (Food Products Standards and Food Additives) III Amendment Regulation 2017. It provides sampling plan and limit of hygiene and safety indicator organisms associated with chilled/ frozen fish, chilled/ frozen crustaceans, chilled/frozen cephalopods, live bivalve mollusks, chilled/frozen bivalves, frozen cooked crustaceans or mollusks, dried or salted fish, thermally processed, fermented, smoked and freeze-dried products, fish mince/surimi, fish pickles, battered/ breaded and convenience fishery products, and powdered fish-based products.

With the availability of a wide array of regulations and standards globally, it becomes a necessity for the FBOs to sell safe and quality food to the consumers. Indian total fish production stands at 12.59 million tons in 2017-18, with the export contribution of 1.38 million tons. The FBOs need to follow the FSSAI regulations as 90% of the fish produced is sold for domestic consumption. They need to fulfill the microbiological criteria, as a majority of the fish (>90%) sold are in raw and chilled forms. The availability of processed fish products in Indian markets is very much limited in the supermarkets of cities and towns. The export of fishery products from India has touched a sum of Rs. 45106.89 crore (7.08 billion$) in 2017-18. The FBOs are therefore paying more attention now-a-days on the quality and the safety of fish and fishery products.

Fish and fishery products fetch a high price in European than in US and Japanese markets. The export requirements for fish and fishery products to the European Union are thus more stringent with several procedures and requirements. The pre-export approvals by the competent authority of importing nations, catch certificate by the Marine Products Export Development Authority (MPEDA), and export certificate by Export Inspection Agency (EIA) are the document requirements for inspection in borders. Besides, there are several product specifications for the export of aquatic products to EU, which include microbiological requirements (EC No. 2703/2005), maximum levels for permitted additives (EC No. 1333/2008), maximum limits for contaminants, including heavy metals, dioxins, PCBs and shellfish toxins, PAHs (EC No. 1881/2006), and maximum levels for therapeutics and antibiotics (EC No. 37/2010). Fish product of aquaculture origin must obtain health certificate as per Council Directive requirement (2006/88/EC). The FBOs are getting their products tested for various quality and safety parameters either in their in-house laboratory or accredited government or private laboratories. The assessment of the quality and safety of aquatic food thus becomes very important. To accomplish this, in-depth knowledge on the analytical procedures to test the aquatic food product quality and safety parameters is required.

Fish and shellfish are good sources of protein, minerals, vitamins, omega-3 fatty acids, and essential amino acids. Fish is a rich source of calcium and phosphorus and also iron, zinc, selenium, iodine, magnesium, and potassium. Fatty fish such as salmon, tuna, sardines, mackerel, anchovies, and trout contain EPA and DHA, as the predominant omega-3 fatty acids. Fish is also a natural source of B-complex vitamins, especially riboflavin (B2), niacin (B3), cyanocobalamin (B12) as well as vitamin D. Fish contains a high protein of around 20% with a high biological value of > 90%. It has a stable composition of essential amino acids with slight deficiencies of methionine and threonine but has an excess of lysine to supplement for cereal grains. A serving of 3 ounce (85g) fish provides about 30-40% of the average daily recommended protein. Several works of literature are available on the chemical composition of widely available food fish of marine and freshwater origin (www.fao.org). In the advent of rapid growth in the aquaculture sector, the feed composition and feeding behaviour of fish largely influence the chemical composition of the aqua cultured fish. The introduction of new processing techniques can lead to the development of a variety of processed fishery products. Different processing methods can affect the nutritive value of aquatic food products, which is yet another aspect of focus in recent years. A thorough understanding of the various methods employed to estimate the chemical constituents in the aquatic food products, is therefore, essential.

CHAPTER 2

ANALYSIS OF
CHEMICAL CONSTITUENTS

Food chemistry is a study of chemical composition, processes, and interactions of all the biological and non-biological components of the food. The biological components include carbohydrates, proteins, fat, vitamins, minerals, and dietary fiber, and the non-biological components consist of the contaminants. The proportion of different chemical components within the food determines the chemical composition. The term "proximate composition" means the proportion of the basic components of the food. The basic components differ from the type of food. Fish and shellfish primarily consist of water, proteins, lipids, and minerals that make up 98% of the total mass. Other minor constituents include carbohydrates, vitamins, and minerals.

Food compositional data (FCD) provides a detailed set of information on the nutritionally important components of foods. FCD also provides values for energy and nutrients including protein, carbohydrates, fat, vitamins, minerals, and dietary fiber. Food compositional databases (FCDB) provide detailed FCD. In 1940, Widdowson released, the first book on "The Composition of Foods" in the United Kingdom. The Food and Agricultural Organization (FAO) has published Food Composition Tables for globally use. Several countries have FCDBs for the foods of their origin. The USA, Japan, and India have DBs for 7500, 2191, and 528 foods, respectively. Some special FCDBs are available to provide details on individual amino acids and vitamin fractions and bioactive compounds, eg. EUROFIReBASIS. The National Institute of Nutrition in Hyderabad released the Indian Food Composition Table (IFCT) in 2017.

The FAO brought out Food Composition Tables for fish and fishery products in 2016 (Fish global database). This database contains a total of 515 food entries that provides nutrient values of fish, crustaceans and mollusks in raw (134 Nos.), cooked (376 Nos.), and processed (5 Nos.) forms along with data on proximate composition, minerals, vitamins, amino acids, and fatty acids. The proximate composition of different fish species is available in databases such as the Fish Base (www.fishbase.org). The IFCT contains fish and shellfish under four groups, namely P, Q, R and S for marine fish (92 Nos.), marine shellfish (8 Nos.), marine mollusks (7 Nos.), and freshwater fish and shellfish (10 Nos), respectively.

2.1. SIMULTANEOUS DETECTION OF MAJOR COMPONENTS

Analysis of each nutrient individually is time-consuming and requires a diverse set of equipment. There are two important non-destructive methods for the simultaneous detection of the major components of fish viz. near-infrared spectroscopy (NIR), and nuclear magnetic resonance spectroscopy (NMR).

2.1.1. Near infrared spectroscopy (NIR)

The NIR-transmittance spectroscopy is a rapid and reliable method for analysis of the major components in fish (Solberg, 1997). Prior to analysis, homogenize the food sample to obtain several sub-samples. Obtain the spectral data of subsamples and average to get the spectral data. Use the spectra data to perform multivariate calibrations against the chemical or physical data. Use the same spectral data to predict water, fat, and protein contents simultaneously. Use an interacting fiber-optic probe to analyze directly on the fish fillet. The fiber-optic probe carries both the incident and reflected radiation. There are several portable instruments available to test the chemical composition in a whole fish as well as live fish (Shimamoto *et al.*, 2003).

2.1.2. Nuclear magnetic resonance spectroscopy (NMR)

The low field NMR is a technique used for the determination of fat, water, and protein in fish. Use a cylindrical sample of 10-40 mm diameter for the analysis. This method is fast, accurate, and easy to perform. A handheld portable NMR mouse is available for usage in live fish within an analysis time of < 20 sec (Veliuylin *et al.,* 2005).

2.2. ANALYSIS OF INDIVIDUAL COMPONENT

There are destructive and non-destructive methods for the analysis of the major individual components such as lipid, protein, and carbohydrates in fish and shellfish.

2.2.1. Lipids

Fish lipid consists of n-3 polyunsaturated fatty acids such as eicosapentaenoic acid (EPA-C20:5n-3) and docosahexaenoic acid (DHA-C22:6n-3). These fatty acids are constituents of different lipid fractions such as triacylglycerols, phospholipids, lysophospholipids, partial glycerides, esters, and free fatty acids. In pelagic fatty fish, the triacylglycerol is the dominant lipid fraction. In lean fish, the phospholipid is the dominant lipid fraction. The other derivatives include fatty acids, sterols, fat-soluble vitamins, and carotenoids. All these altogether constitute the "total lipids". Destructive and non-destructive methods are used to determine the lipid content in fish and shelfish. Destructive methods include organic solvent extraction and microwave drying, while non-destructive methods include fat meters, NIR, and NMR.

2.2.1.1. Destructive chemical methods

The chemical methods involve solvent extraction followed by gravimetric determination. The yield is dependent on the solvent system. A combination of polar and non-polar solvent is employed to extract the total lipids in fish, for instance, chloroform, methanol, and water. The methanol penetrates the fish tissue while the chloroform dissolves the fat out of the fish matrix.

2.2.1.1.1. Bligh and Dyer method

This method determines the lipids in fish muscle (Bligh and Dyer, 1959). Sodium chloride and electrolyzed cathode water used to get a higher yield of lipid in the modified method. The extraction allows the characterization of lipids, the different lipid classes, the lipid oxidation products, and the fatty acid composition. The method gives an average recovery of 95% of lipid. The method is relatively time-consuming, requires laboratory facilities, and involves toxic solvents.

Reagents

1. Chloroform and methanol (2: 1)
2. Chloroform
3. Potassium chloride (KCl), 0.88%

Procedure

Homogenize 10±1 g of wet fish sample with 10 ml of distilled water and transfer the pulp into a conical flask. To the pulp, add 30 ml of chloroform: methanol (2:1), mix well and allow to stay at room temperature for 10-30 min, preferably in dark. Add 20 ml of chloroform and 20 ml of distilled water, mix well and filter the homogenate through the strainer first and then through the filter paper. To remove the impurities, add 10 ml of 0.88% KCl to the filtrate. Transfer the entire content to a separating funnel and allow the two layers to separate. Remove the lower layer containing the lipid carefully into a pre-weighed beaker. Evaporate the chloroform solvent by placing it in a water bath to concentrate the lipid. Place the beaker in a desiccator and weigh to determine the lipid content.

Calculation

$$\text{Total lipid (\%)} = \frac{\text{Wt. of beaker with fat} - \text{Wt. of empty beaker}}{\text{Wt. of wet fish}} \times 100$$

2.2.1.1.2. Folch method

This method is applied when more exhaustive recoveries of lipids are required (Folch *et al.*1957). The average recovery by this method is 95-99%.

Reagents

1. Chloroform and methanol (2:1)
2. Potassium chloride (KCl), 0.74%
3. Chloroform, methanol, and water (3:48:47)

Procedure

Homogenize 10±1 g of wet sample with 30 ml of chloroform: methanol (2:1) and allow to stay in the room temperature for 10 min in dark. Filter the entire content first using a strainer and then through a filter paper. To the residue, add 30 ml of chloroform: methanol (2:1) again to extract the lipid and filter as described above. Combine the filtrates. Add 10 ml of 0.74% KCl and shake well. Transfer the entire content to a separating funnel and allow the two layers to separate. Remove the lower chloroform layer and transfer it to another separating funnel and then, add 10 ml of chloroform: methanol: water (3: 48: 47) mixture and shake well. Allow the two layers to separate again. Remove the lower chloroform layer in a pre-weighed beaker and evaporate to dryness in a water bath. Place the beaker containing the lipid in a desiccator and weigh to determine the lipid content.

Calculation

$$\text{Total lipid (\%)} = \frac{\text{Wt. of beaker with fat} - \text{Wt. of beaker}}{\text{Wt. of wet fish}} \times 100$$

2.2.1.1.3. Automatic solvent extraction method

The automatic solvent extraction technique developed by Soxhlet in 1879 is the most widely used method for the determination of total lipids in fish (AOAC, 1990). Before the extraction, lyophilize or dry the sample and then remove the solvents before gravimetric determination. Petroleum ether and diethyl ether are the most common solvents for the extraction. The traditional methods are time-consuming, and presently there are more rapid methods available based on the same principle with commercial instrumentation such as SoxTec and SocPlus. A new microwave integrated Soxhlet apparatus can perform the analysis in less than an hour (Virot *et al.,* 2007).

Procedure

Weigh accurately 2±0.2 g dry sample, and transfer into a thimble. Place the thimble in the holder provided with the apparatus. Pre-weigh the extraction flask and assemble in the apparatus below the thimble on the heating platform. Fill the condenser with the petroleum ether until it overflows and fills up to 2/3 volume in the extraction flask. Turn 'ON' the instrument and extract the lipids for about 2 h at 200°C without any interruption. After extraction, remove the thimble first and re-assemble the extraction flask without thimble to remove the excess ether into the condenser flask for reuse. Evaporate the residual ether in the flask, by placing it in a heating mantle for a while. Cool the flask and place it in a desiccator. Weigh the flask to determine the lipid content.

Calculation

$$\text{Crude lipid (\%)} = \frac{\text{Wt. of flask with fat} - \text{Wt. of flask}}{\text{Wt. of dried fish}} \times 100$$

2.2.1.2. Non- destructive instrumental methods

2.2.1.2.1. Microwave drying

Microwave drying is a simple and inexpensive method that indirectly calculates the lipid content from the analysis of water (Vogt *et al.*, 2002). The principle is the reversed inter-correlation between water and lipid contents in fish. The lipid content is calculated by the formula: lipid content, % = 80% - water content, %.

2.2.1.2.2. Fat meter

The portable fat meter determines the total lipid in fish and shellfish (Kent, 1990). Irradiate the sample by microwaves with a microwave strip and then, measure the water by the dielectric properties, and calculate the lipid content. Calibrate these instruments for a range of fish species.

2.2.2. Proteins

Fish protein has high nutritive value with the right balance of essential and non-essential amino acids. They are highly digestible with raw fish having 90-98% digestibility, and shellfish about 85%. Fish contains about 11-24% crude protein depending on species, nutritional conditions, and type of muscle. There are various methods for protein determination in foods like Dumas, Nessler's reagent, Biuret, Kjeldahl, Folin-Ciocalteau, and Dye binding. There are direct and indirect methods for the determination of total proteins in fish.

2.2.2.1. Indirect methods

Protein content is determined by the analysis of total nitrogen (N) first and then multiplying it by a factor specific for fish (Greenfield and Southgate, 2003). The total nitrogen is determined either by Kjeldahl or Dumas methods.

2.2.2.1.1. Kjeldahl method

This method is first published by Kjeldahl in 1883 and later modified by several authors extensively. The method involves sample digestion, neutralization, distillation to trap ammonia, and titration. This method is adopted as a reference method by many national and international organizations. The major advantage of this method is the production of accurate results, and the disadvantage is the use of hazardous and toxic chemicals. Several instruments developed for automated analysis include Kjel-Foss, Kjel-Tec, and KjelPlus.

The principle behind the method is that the sample is first hydrolyzed with the acid during the digestion step and the nitrogen evolved is converted to ammonium sulfate. The total nitrogen that is transformed into ammonium ions is distilled as ammonia in the presence of alkali, then trapped in boric acid as a complex and quantified back by titration with acid to determine the total nitrogen.

Digestion step

Sample is digested with acid at this step to yield the digested product, ammonium sulfate

$$\text{Organic N} + H_2SO_4 \rightarrow (NH_4)_2SO_4 + H_2O + CO_2 + \text{other sample matrix byproducts}$$

Distillation step

To the digestion product, excess alkali is added to convert NH_4 to NH_3 during distillation.

Ammonium sulfate		Ammonia gas heat	
$(NH_4)_2SO_4 + 2NaOH$	$\rightarrow$	$2NH_3$	$+ Na_2SO_4 + 2H_2O$

Boric acid is used as the receiving solution to capture the ammonia gas forming an ammonium-borate complex. The color of the receiving solution changes with the collection of ammonia to green, if mixed indicator is used.

Ammonia gas		Boric acid	Ammonium-borate complex		Excess boric acid
NH_3	$+$	H_3BO_3	$\rightarrow$ $NH_4 + H_2BO^-_3$	$+$	H_3BO_3

Titration step

The nitrogen present in the ammonium-borate complex is determined by titrating back with standard mineral acid. The color of the solution reverses back to red color.

Ammonium-borate complex	Hydrochloric acid	Ammonium sulfate	Boric acid
$NH_4^+ H_2BO_3^-$ +	HCl $\rightarrow$	NH_4Cl	+ H_3BO_3

(colour change occurs in reverse)

Calculate the protein content by multiplication with the Kjeldahl conversion factor of 6.25. The conversion factor is arrived based on the average nitrogen content of 16% in the animal protein. In fish, this conversion factor ranges from 5.43 to 5.82. However, the factor of 6.25 is considered acceptable, as it is used for more than 75 years.

Reagents

1. Conc.H_2SO_4
2. Digestion mixture: Copper sulfate – 0.1 g and Potassium sulfate - 2.5 g for each sample
3. Sodium hydroxide, 40%
4. Boric acid, 4 %
5. Mixed Indicator: Methyl red - 0.16 g and bromocresol green - 80 mg in 100 ml of 95 % ethanol.
6. Standard H_2SO_4, 0.1 N

Procedure

Digestion: Weigh the sample (0.2 g dry weight or 0.5 g wet weight) and transfer it into the digestion tube. Add a pinch of the digestion mixture along with 10 ml of concentrated H_2SO_4. Place the digestion tube in the sample digester and fit the exhaust manifold on top of it. Switch 'ON' the power and maintain the heat between 300 - 400°C. Turn 'ON' the cooling water during digestion. Allow the sample to digest until the solution becomes colorless or light green. Remove the digestion tube from the digester, cool the content, and dilute to 100 ml in a volumetric flask with distilled water. Digest a blank without sample but with 5 ml of distilled water in a similar way to determine the presence of any residual nitrogenous substances in the solvent.

Distillation: The automatic N_2 analyzer operated as described in the manual. Start the process, when the "ready" displays. Place the empty digestion tube with little distilled water in the distillation unit, and press the "dilution key" on the control unit. On the other side, keep the empty receiver flask on the platform in the distilling unit. Press the "process" key, and run for 5–10 min to clean up the system. Start the actual experiment after this process is over. In the instrument program, set the "alkali" to 10 ml, and "delay time" to 8 – 10 min. Check the availability of distilled water, and cooling water flow. Take 5 ml of the diluted digested sample in the digestion tube and place it in the position of the distilling unit. Place the receiver flask on the platform of the distilling unit with 10 ml of boric acid containing one or two drops of mixed indicator. Close the safety door during the operation. Set the "time" for alkali (4 sec is sufficient for 10 ml of alkali), and then set the "min" for the process (8 – 10 min is sufficient for one distillation). Press the "run" key to start the distillation process. After the distillation is complete, the steam lamp shows "steady" and a "beep" sound is heard. Remove the digestion tube with the residue. Remove the receiver flask with the distillate. Continue the operation with the next sample as well as blank.

Titration: Titrate the content in the receiver flask with standard hydrochloric acid until the appearance of the green color endpoint. Obtain the titration volume of the sample by subtracting the titration volume of the blank.

Calculation

$$\text{Total crude protein, \%} = \frac{1.4 \text{ x V x } T_1 - T_2 \times 100 \times 100 \times 6.25}{W \times 5 \times 1000}$$

where, V – volume of the sample, ml; W – weight of the sample, g; T_1 – titre value of sample, ml; T_2 – titre value of blank, ml; 1000 – mg to g conversion; 100 – made up volume;100 – percentage conversion; 1.4 – equivalent N_2 for 0.1 N H_2SO_4; 6.25 – conversion factor

2.2.2.1.2. Dumas method

In the method first published in 1831, all the total nitrogen is converted to nitrogen through combustion using a nitrogen element analyzer (Dumas, 1831). Put the sample in a furnace of 950-1050°C, purge free of atmospheric gas, and then fill it with pure oxygen. The gas mixture consists of CO_2, SO_2, O_2. Cool the sample with water and remove. The NO_2 gets reduced to N_2. Measure the reading with a thermal conductivity meter. Calibrate the combustion method first with the Kjeldahl method. The method is quick, cheap, and simple but gives higher nitrogen values. For fish, Kjeldahl-N to Dumas-N ratio is 0.80.

2.2.2.2. Direct methods

Proteins are composed of amino acids linked together by peptide bonds. Therefore, the quantification of these amino acids gives the most accurate values for protein.

2.2.2.2.1. HPLC and IEF methods

Acid hydrolysis followed by amino acid quantification by High Performance Liquid Chromatography (HPLC) and traditional Iso Electric Focussing (IEF) are the direct and specific methods of protein determination. In IEF method, the amino acids are derivatized post column using ninhydrin ortho-phthalaldehyde (OPA).

In HPLC method, the common derivatization reagents are OPA and 9-fluorenylmethyl chloroformate (FMOC), which are used in combination with 2 mercaptoethanol, ethanthiol and 3-mercaptopropionic acid. A combination of OPA and FMOC helps to detect primary and secondary amino acids (Heems *et al.*, 1998). Pre-and post-column derivatization can be used. The mobile phases are methanol and acetonitrile. Hydroxyproline found in connective collagenous tissue is quantified through derivatization with 7-chloro-4-nitrobenzo-2-oxa-1,3-diazole and N_2-(5-fluoro-2,4-dinitrophenyl)-L-valine amide. Destruction of tryptophan during acid hydrolysis can be omitted by replacing with methane sulphonic acid containing 3-(2-aminoethyl) indole. Detection of cysteine require special procedures during extraction using iodoacetic acid or 3,3'-dithiodipropionic acid. The quantification of the individual amino acids is based on standard amino acids, their response and molecular weights and use of an internal analytical standard such as alpha-butyric acid (ABA). The total protein values are calculated as the sum of all the amino acids.

2.2.2.2.1.1. UPLC method for amino acid analysis

The protocol mentioned below gives the analysis of amino acid composition using H Class-Ultra Performance Liquid Chromatography (UHPLC) as described by Waters Corporation Pvt. Ltd. The derivalization reagent in AccQ•Tag^{TM} reagent, 6-aminoquinolyl-N-hydroxy succinimidyl carbamate.

Sample hydrolysis: Take 20±0.02 mg of sample in a test tube and add 20 ml of 6 N HCl. Flush the tubes with nitrogen gas, seal the tube, and place it in a hot air oven set at 110°C for 24 h for hydrolysis. Filter the sample through Whatman No.1 filter paper to remove the un-hydrolyzed debris and neutralize to pH 7.0 using the same volume of 6 N NaOH. Filter the neutralized sample through a syringe filter (0.2µm).

Derivatization: Derivatize the sample using the AccQ•Tag™ Ultra pre-column derivatization kit. Mix 10ml of filtrate with 70 ml of AccQ•Tag™ Ultra borate buffer and 20ml of reconstituted AccQ•Tag™ Ultra derivatization reagent for derivatization and then transfer the sample into autosampler vials of UHPLC.

UHPLC analysis: In this method, a Waters ACQUITY-UHPLC fitted with AccQ Tag ULTRA C18, 1.7mm, 2.1x100 mm column used for the analysis of amino acids. The mobile phase consists of four solvents, A - Eluent A, B – MilliQ water : Eluent B (90:10, v/v), C - MilliQ water, D – Eluent B. A gradient elution program for the chromatographic separation consists of the following: Initial-0.29 min, A 2%, C 98%; 0.29-5.49 min, A 9%, B 80%, C 11%; 5.49-7.30 min, A 8%, B 15.6%, D 18.5%; 7.30-7.69 min, A 7.8%, C 70.9%, D 21.3%; 7.69-8.59 min, A 4%, C 36.3%, D 59.7%;8.59-10.20 min, A 2%, C 98% at 0.7 ml/min flow rate. Set the column temperature as 55°C, the injection volume as 2µl, and the tunable UV detector at 260nm. Perform the data analysis by Empower 2 software. Calculate the concentration of individual amino acids in the sample with the help of an authentic standard run alongside in the same condition. Express the results in mg/100g. Calculate the total protein content as the sum of all the amino acids.

2.2.2.2.2. Spectrophotometric methods

Spectrophotometric methods determine the protein content in fish and shellfish. Some methods depend on the ability of proteins to absorb the light, whereas, in other methods, the proteins modified either chemically or physically to absorb the light. The variation in the amino acid composition gives different results for the absolute protein concentration. The methods categorized into two groups are dye-binding reaction and redox reaction with proteins. The redox reaction-based assays include Biuret, Lowry, and Bicinchoninic acid (BCA). The Bradford method depends on the dye-binding reaction.

2.2.2.2.2.1. Biuret method

Biuret method depends on the formation of complexes between copper salts and peptide bonds under alkaline conditions (Noll *et al.*, 1974). The method is simple and inexpensive but is not very sensitive to measure concentrations between 1 to 10 mg/ml (Sozgen, 2006).

In this reaction, Cu (II) is first reduced to Cu (I) in an alkaline medium, which then binds to protein forming Cu(I)-peptide complex producing purplish violet colour, having an absorption maximum at 540nm.

Reagents

1. Trichloroacetic acid TCA, 15%
2. Sodium hydroxide NaOH, 1 N
3. Sodium hydroxide NaOH, 10%
4. **Biuret reagent:** Dissolve 1.5g of copper sulphate and 6g of sodium potassium tartrate in 500 ml of distilled water. Add 300 ml of 10% sodium hydroxide solution and make up to 1000 ml with distilled water.
5. **Protein standard solution (5mg/ml):** Dissolve 125 mg of bovine serum albumin crystal in little amount of 1N NaOH in a volumetric flask and make up to 25 ml with 1N NaOH.

Procedure

Weigh 5±0.5 g of the sample and homogenize with 5 ml of distilled water and 5 ml of 15% TCA. Centrifuge the homogenate at 5000xg for 10 min. Discard the supernatant and re-dissolve the precipitate in 5 ml of 1N NaOH. Transfer 0.5 and 1.0 ml of the sample solution into two test tubes, and make up the volume to 2.0 ml with 1N NaOH. Transfer 0.4, 0.8, 1.2, 1.6, and 2.0 ml of the protein standards into a series of test tubes, and make up the volume to 2.0 ml with 1N NaOH. Prepare a blank with 2.0 ml of 1N NaOH in another test tube. Add 8 ml of Biuret reagent to all the tubes and leave them at room temperature for 30 min. Measure the absorbance (O.D) at 540 nm in a spectrophotometer. Draw a standard graph by plotting the concentration of protein in the X-axis and O.D on the Y-axis. Determine the concentration of the protein from the graph. The software in the spectrophotometer automatically computes the values and gives the concentration of the protein in the sample directly.

Calculation

$$\text{Total protein, \%} = \frac{C \times D.F \times 100}{W \times 1000}$$

where, C – concentration of protein, mg; W – weight of fish, g; D.F – dilution factor (10 ml); 100 – percentage conversion; 1000 – mg to g conversion

2.2.2.2.2.2. Lowry method

Lowry method is the most popular method used to measure protein concentration in food (Lowry *et al.,* 1951). The protein determined by this method is comparable to those obtained by the Kjeldahl method. This method is simple, sensitive, precise and suitable for protein extracts such as actomyosin. The method is sensitive at low concentrations of protein, ranging from 0.1 to 2.0 mg of protein/ml. The disadvantages are the interfering substances such as some amino acid derivatives, reducing sugars, drugs, lipids, nucleic acids, sulfhydryl groups, salts, and several common buffers (Dunn, 1992).

The principle behind this method lies with the reaction of peptide nitrogen with Cu (II) ions under alkaline medium (pH 10-10.5) to produce Cu(I) and later Cu(I) reacts with the Folin reagent. Amino acids such as tyrosine and tryptophan reduce the phospho molybdic phospho tungstic components to hetero polymolybdenum/tungsten that gives a strong blue colour having an absorbance maximum at 660 nm.

Reagents

1. Reagent A: Dissolve 2% sodium carbonate (Na_2CO_3) in 0.1N NaOH

2. Reagent B: Dissolve 0.5% copper sulphate ($CuSO_4$) in 1% sodium potassium tartrate

3. Reagent C(alkaline copper compound): Mix 50 ml of A and 1 ml of B, prior to use

4. Folin – Ciocalteau reagent: Readymade solution available.

5. **Protein standard solution**

    Stock (1mg/ml): Dissolve 50 mg of bovine serum albumin crystal in little amount of 1N NaOH in a volumetric flask and make up to 50 ml with 1N NaOH.

    Working (200µg/ml): Dilute 10 ml of the stock solution to 50 ml with distilled water in a volumetric flask

6. **Tris-HCl buffer, 50 mM (pH-7.5):** Dissolve 605 mg of Tris in 50 ml of distilled water. Adjust the pH to 7.5 with 0.05N HCl and make up the volume to 100 ml with distilled water.

Procedure

Weigh 5±0.5 g of sample and homogenize using 5 ml of the Tris-HCl buffer and centrifuge at 5000xg for 10 min. Transfer 0.1 and 0.2 ml of the extract into the test tubes, and make up the volume to 1.0 ml with distilled water. Transfer 0.2, 0.4, 0.6, 0.8, and 1.0 ml of protein working standards into a series of test tubes, and make up the volume to 1.0 ml with distilled water. Prepare a blank with 1.0 ml of distilled water. Add 5 ml of reagent C to all the test tubes, mix well and allow to stand for 10 min. Add 0.5 ml of Folin–Ciocalteau reagent to all the test tubes, mix well and incubate the tubes in the dark for 30 min at room temperature, for the color development. Measure the absorbance (O.D) at 660 nm in a spectrophotometer. Draw a standard graph by plotting the concentration of protein in the X-axis and O.D on the Y-axis. Determine the concentration of the protein from the graph.

Calculation

$$\text{Protein, } \% = \frac{C \times D.F \times 100}{W \times 1000}$$

where, C – concentration of protein, mg; W – weight of fish, g; D.F – dilution factor (5 ml); 100 – percentage conversion; 1000 – mg to g conversion

2.2.2.2.2.3. Bicinchoninic acid method

This assay determines the total protein content in an easy way at concentrations from 0.5 µg/mL to 1.5 mg/mL (Walker, 2002). The principle behind is the peptide bond in protein reduces Cu^{2+} ions from the Cu(II) sulfate to Cu(I), which is proportional to the amount of protein in the solution. The molecule of bicinchoninic acid chelates with each Cu (I) ion at a higher temperature (40-60°C) to give an intense purple colour complex having an absorbance maximum at 562 nm. The bicinchoninic Cu (I) complex formation depends on the presence of cysteine, tyrosine, and tryptophan side chains.

2.2.2.2.2.4. Bradford method

Bradford method is a dye-binding reaction-based assay, which is simple, sensitive, and requires a shorter analysis time of 5 min (Bradford, 1976). This assay is less affected by reagents and non-protein components from biological samples, except excess SDS and buffers.

The principle of this method is based on the absorbance shift of the dye, Coomassie Brilliant Blue G-250, which exists in three forms – anionic (blue), neutral (green) a cationic (red). The positively charged amino acid residues in proteins react with Coomassie Brilliant Blue G-250 under acidic condition and form a strong, non-covalent insoluble complex with carbonyl group of the protein by van der Waals force and with amino group through electrostatic interaction. This complex converts red form of the dye into blue form having an absorbance maximum at 595 nm. The amount of complex formed is a measure of the protein concentration.

Reagents

1. Coomassie Brilliant Blue 1
2. NaCl, 0.15 M
3. **Bovine serum albumin (BSA):** Prepare a series of standards diluted with 0.15 M NaCl to get final concentrations of 250, 500, 750 and 1500 µg/mL.
4. **Phosphate buffer, pH 7.0:** Dissolve 3.403 g of anhydrous monopotassium dihydrogen phosphate (KH_2PO_4) and 4.355 g of anhydrous dipotassium hydrogen phosphate (K_2HPO_4) in 100 ml of distilled water. Combine the solution and dilute to 1000 ml with distilled water

Procedure

Weigh 0.5±0.05 g of the sample and homogenize using 5 ml of the phosphate buffer. Centrifuge the homogenate at 5000xg for 10 min. Transfer 0.1 ml of the supernatant into a test tube, and make up the volume to 1.0 ml with the phosphate buffer. Transfer 0.1 ml of each protein standard into the test tube and make up the volume to 1.0 ml with the buffer. Prepare a blank with 1.0 ml of 0.15 M NaCl in a test tube. To all the test tubes, add 5.0 µl of Coomassie Blue solution, mix well, and allow to stand for 5 min at room temperature. Measure the absorbance (O.D) at 595 nm in a spectrophotometer. Draw a standard graph by plotting the concentration of protein in the X-axis and O.D on the Y-axis. Determine the concentration of the protein from the graph.

Calculation

$$\text{Protein, \%} = \frac{C \times D.F \times 100}{W \times 10^6}$$

where, C – concentration of protein, mg; W – weight of fish, g; D.F – dilution factor (5 ml); 100 – percentage conversion; 10^6– µg to g conversion

2.2.2.2.2.5. Simple spectrophotometer analysis

The protein in solutions can be quantified by near or far UV absorbance simple spectrophotometric analysis (Aitken and Learmonth, 2002). Absorption in the near UV depends more on the content of tyrosine and tryptophan, and less on the amount of phenylalanine and disulfide bonds. The absorbance method is simple, sensitive, needs no reagents, and the sample is recoverable. Crude protein extracts or individual fractions of proteins measured at 280 nm. Disadvantage at this wavelength includes interference by other components such as nucleic acid, which absorbs in the same wavelength region.

Far UV absorption used to determine the protein content, as peptide bonds absorb in this area with the absorption maximum at 190 nm. Different proteins give a small variation in absorbance and considered as accurate for protein determination. Oxygen also absorbs at this wavelength, and hence to avoid interference, it is advisable to measure at 205 nm. The presence of carbohydrates, salts, lipids, amides, phosphates, and detergents interferes at this wavelength.

Comparison of methods to determine protein concentrations

Method	Range (µg)
Kjeldahl	500 - 30,000
Biuret	1,000 – 10,000
Lowry	10 - 300
Bradford	20 - 140
Bicinchoninic acid	1 - 50
Absorption at 280 nm	100 - 300

2.2.3. Carbohydrates

Carbohydrates are classified into three broad groups: sugars (mono and disaccharides), oligosaccharides (3 to 9 monosaccharides), and polysaccharides (>9 monosaccharides). The carbohydrates in the fish muscle are quite low, while some marine invertebrates possess high contents. Sub-cuticular tissue of spiny lobster and blue crabs contain 10.2% and 12.5% of total sugars, respectively, with the high amounts of glucose followed by galactose and mannose.

2.2.3.1. Phenol-sulfuric acid method

Total carbohydrates is determined by the phenol-sulfuric acid method of Dubois *et al.* (1956) in fish and shellfish. The method measures all classes of carbohydrates, including mono, di, oligo, and polysaccharides. This method depends on the hydrolysis of polysaccharides. The carbohydrates absorb at different maximum wavelengths and differ in their ability to form the chromogens. This method hence does not measure all sugar molecules accurately. Measurement of glucose at 490 nm leads to an underestimation of monosaccharides other than glucose. Nevertheless, this method gives an estimate of total carbohydrates in tissue that contain 10% or more of hexose polymers.

The principle of this method is that the concentrated sulphuric acid breaks down any polysaccharides, oligosaccharides, and disaccharides into monosaccharides. The pentoses are then dehydrated to furfural and hexoses to hydroxymethyl furfural. These compounds react with phenol to produce a green colour product having an absorption maximum at 490 nm. For the products high in hexose sugars, glucose is commonly used to create the standard curve.

Reagents

1. Phenol, 5% (redistilled)
2. Sulphuric acid, 96%
3. HCl, 2.5 N
4. Sodium carbonate
5. **Standard Glucose Solution**

    Stock (1 mg/ml) – Dissolve 100mg of D-glucose in a volumetric flask and make up to 100ml with distilled water.

    Working (100µg/ml)– Dilute 10ml of stock solution in a volumetric flask and make up to 100ml with distilled water.

Procedure

Weigh 0.1 ± 0.01 g of the sample in a boiling tube with 5ml of 2.5 N HCl. Place the tubes in a boiling water bath for hydrolysis and then cool. Neutralize the solution with solid sodium carbonate until the effervescence ceases, and make up the volume to 100ml with distilled water, and then centrifuge at 5000 rpm for 5 min. Transfer 0.2, 0.4, 0.6, 0.8, and 1.0 ml of the working standards into a series of test tubes, and make up the volume to 1.0 ml with distilled water. Transfer 0.1 and 0.2ml of the sample extracts into two separate test tubes and make up the volume to 1.0 ml with distilled water. Prepare a blank with 1.0 ml of distilled water. Add 1.0 ml of phenol solution to all the test tubes followed by 5ml of 96% sulphuric acid and shake well. Leave the tubes for 10min, and then place them in a water bath set at 25-30°C for 20min, for color development. Read the absorbance (O.D) at 490 nm in a spectrophotometer. Draw a standard graph by plotting the concentration of glucose in the X-axis and O.D on the Y-axis. Determine the concentration of the glucose in the sample from the graph.

Calculation

$$\text{Total carbohydrates, } \% = \frac{C \times D.F \times 100}{W \times 10^6}$$

where, C – conc. of glucose, mg; W – weight of fish, g; $D.F$ – dilution factor (100 ml); 100 –percent conversion; 10^6 – μg to g conversion

2.2.3.2. Anthrone method

Small amount of glycogen present in the tissues is analyzed by Anthrone method (Carroll *et al.*, 1955).

Carbohydrates are first hydrolyzed into simple sugars using dilute hydrochloric acid or trichloro acetic acid (TCA). Glucose is dehydrated to hydroxymethyl furfural in hot acidic medium, which then reacts with anthrone to form a green colored product having an absorption maximum at 630 nm.

Reagents

1. Trichloro acetic acid (TCA), 5%

2. **Anthrone reagent:** Dissolve 50 mg of anthrone in 100 ml of 70% cold sulfuric acid

3. **Glucose standard:**

   Stock standard (1mg/ml): Dissolve 100 mg of glucose in 100 ml of distilled water in a volumetric flask.

   Working standard (100µg/ml): Pipette out 5 ml of stock and make up to 50 ml with distilled water in a volumetric flask.

Procedure

Weigh 2±0.2 g of the sample and homogenize with 10 ml of distilled water. Add10 ml of 5% TCA solution, mix well, centrifuge at 5000xg for 10 min and filter. Transfer 0.5 and 1.0 ml of the filtrate into two test tubes and make up the volume to 1.0 ml with distilled water. Transfer 0.2, 0.4, 0.6, 0.8, and 1.0 ml of the working standards into a series of test tubes and make up the volume to 1.0 ml with distilled water. Prepare a blank with 1.0 ml of distilled water tube. Add 10 ml of anthrone reagent to all the test tubes and place them in a boiling water bath for 10-15 min, for color development. Cool the tubes for 30 min at room temperature. Measure the absorbance (O.D) at 620 nm in a spectrophotometer. Draw a standard graph by plotting the concentration of glucose on the X-axis and O.D on the Y-axis. Determine the concentration of the glucose in the sample from the graph.

Calculation

$$\text{Total free carbohydrates, } \% = \frac{C \times D.F \times 100}{W \times 10^6}$$

where, C – conc. of glucose, mg; W – weight of fish, g; D.F – dilution factor (30 ml); 100 –percent conversion; 10^6 – µg to g conversion

Protein bound carbohydrates

Carbohydrates exist as free sugars and bound sugars. For the estimation of protein-bound carbohydrates, add 5 ml of 1 N sulfuric acid, to the TCA precipitate taken in a stopper test tube, and stopper the tube. Place it in an oven set at 100°C for 12-14 h for hydrolysis. Take 1 ml of the hydrolyzed sample in another test tube and add 10 ml of anthrone reagent. Determine the carbohydrate content as done for free sugars.

2.2.3.3. Difference method

In difference method, calculate the carbohydrate and express as total carbohydrates by difference, which is the remainder after subtraction of moisture, crude protein, total fat, and ash, but this also includes fiber, if present.

2.2.4. Crude fiber

Crude fiber consists largely of cellulose and lignin (97%), and some mineral matter. In marine animals, the crude fiber is absent. The marine algae, as well as seaweeds, contain crude fiber. They are consumed in some parts of the world as human food. In addition, aquaculture feed used for farming fish and shellfish contain crude fiber, as they are generally prepared with vegetable-based ingredients.

Oxidative hydrolytic degradation of cellulose and degradation of lignin occurs during acid and subsequent alkali treatment of samples. The residue obtained after filtration is weighed, incinerated, cooled and weighed. The loss in weight gives the crude fiber content.

Reagents

1. Hydrochloric acid (HCl), 1%.
2. Sulfuric acid, (H_2SO_4), 10% and 1.25%
3. Sodium hydroxide (NaOH), 10% and 1.25%
4. Petroleum ether (40-60°C)
5. Ethanol (95-96%)
6. Acetone
7. Antifoam: 2% silicon antifoam in carbon tetrachloride

Procedure

Weigh 1±0.1 g of fat-free sample in a conical flask, add 200ml of 1.25% H_2SO_4 with few drops of antifoam and bring to the boiling point. Boil the content gently for exactly 30 min under the condenser. Filter the content of flask through a wet Whatman No. 1 filter paper. Transfer the sample back into the original flask by washing with 200 ml of 1.25% sodium hydroxide and boil for exactly 30 min. Filter the content of flask again through a wet Whatman No. 1 filter paper. Transfer all the insoluble matters in the filter paper into the pre-weighed sintered crucible by washing with boiling water, and then with 1% HCl, and again with boiling water. Wash the insoluble matter twice with alcohol and thrice with acetone. Dry the matter initially at 100°C in a hot air oven, weigh and finally ash in a muffle furnace at 550°C for 1h. Cool the content in the crucible in a desiccator and re-weigh again.

Calculation

$$\text{Crude fiber content, \%} = \frac{(W_2 - W_3)}{W_1} \times 100$$

where, W_1 – weight of sample, g; W_2 – weight of insoluble matter, g; W_3 – weight of ash, g

2.2.5. Water

The water content of fish generally ranges from 60 -80%, with the exception that Bombay duck having the highest water content of about 90%. In mollusks and crustaceans, the water content is high compared to finfish and elasmobranches. Water content has got a correlation with the water activity, which in turn influences the growth of spoilage microorganisms in foods.

Simple drying methods determine the water content in fish and shellfish. The conventional oven dries the sample at 105°C for 12 h (AOAC, 1990). To ensure complete drying, initially, dry the sample to a constant weight. Other methods include drying at 101°C for 24 h by conventional oven and 70°C for 24 h by vacuum oven. Infrared and microwave oven analysis time is 1-2 h. The new nondestructive methods, including NIR/NIT, and NMR, are employed for the fast determination of water. Low field NMR can distinguish between free and bound water.

2.2.5.1. Hot air oven method

A known quantity of sample is taken and the water content is removed by heating in a hot air oven. The difference in the weight gives the water content in terms of percentage.

Procedure

Pre-weigh a glass dish previously dried in an oven. Weigh accurately 10±1 g of the sample and distribute the contents evenly in the dish. Dry the sample in the dish in a hot air oven maintained at 100°C ± 2°C for 12 h. After drying, cool the dish in a desiccator and weigh it again. Record the final weight. Determine the water content by the following formula:

$$\text{Water content, } \% = \frac{(W_1 - W_2)}{W_1} \times 100$$

where, W_1 – weight of sample, g; W_2 – weight of the sample after during, g

2.2.5.2. Infrared moisture balance

Infrared moisture balance measures the water content of foods that change its chemical structure by losing water (Christie *et al.*, 1985). It measures the water content in a product containing higher amounts of carbohydrates as well as substances that quickly reabsorb moisture after drying. Infrared lamp does the heating. Drying and weighing performed simultaneously. The balance scale is divided directly into a moisture percentage from 0 to 100% in 0.1% divisions.

Procedure

First, turn 'ON' the scale lamp in the instrument. Turn the scale by adjusting knob until the 100% coincides with the center zero of the Vernier scale. Raise the lamp housing carefully and place the test material on the sample pan until the scale reading is 10g. This reading gives the initial weight of the material. Lower the lamp housing and turn 'ON' the infra-red lamp. Adjust the temperature and the timer in the instrument for heating, based on the material requirement. Note down the reduction in the water content during the operation. When heating is complete, note down the reading of moisture percentage and estimate the water content.

2.2.6. Energy

The energy content of food is in kilocalories (kcal) and kilojoules (kJ), which have a conversion factor of 1 kcal = 4.184 kJ. Aquatic food shows the variable composition of proteins and fat, and hence, the energy content is dependent on their distribution.

2.2.6.1. Direct measurement

Gross energy in food is determined directly by a bomb calorimeter (micro or macro methods). This method involves the burning of food with oxygen in an insulated container of constant volume (Miller and Payne, 1959). Heat gets absorbed in water. The energy is measured from the mass of water, its temperature rise, and its specific heat. Dichromatic wet oxidation with potassium chromate is the direct method for energy determination. This method gives rise to slightly lower energy values than that measured by the bomb calorimetric method.

2.2.6.2. Indirect measurement

Direct measurement of gross energy is not equivalent to energy requirements, because the body requires energy to maintain the normal process of life and to meet the demands of activity and growth. Hence, the measurement of metabolizable food energy accounts to the energy in food remaining after losses through the feces, gas, urea, and body surface. Energy intake is the sum of metabolizable energy provided by the available carbohydrates, fat, protein, and alcohol of the digested food.

The energy released by the oxidation of protein, fat, and carbohydrates is the basis for sets of conversion factors. Atwater general factor is the foundation for the most frequently used system for energy conversion (Atwater and Woods, 1896). This system originated from combustion with adjustments for losses in digestion, absorption, and excretion of urea. Atwater general energy conversion values are 4.0 kcal/g for proteins, 9.0 kcal/g for lipids, and 4.0 kcal/g for carbohydrates (more specifically 3.75 kcal/g for monosaccharides and 4.20 kcal/ g for polysaccharides). Available carbohydrates include the sum of glucose, fructose, sucrose, maltose, lactose, dextrin, and starches in the diet. The conversion factors used for the calculation of the energy by the different countries are given, although minor differences exist.

Components	Conversion factor (Kcal/g)	Conversion factor (kJ/g)
Fat	9.00	37
Protein	4.00	17
Available carbohydrates	3.75	16
Starch	4.10	–
Saccharose	3.90	–
Glucose, Fructose	3.75	16
Alcohol	7.00	29

Source: UK Ministry of Agriculture, Fisheries and Food

Calculation

Let, Protein content, % = P; Fat content, %= F; Available carbohydrate, % = C

Then,

Caloric value (Kcal/100g) = P x 4.00 + F x 9.00 + C x 3.75

Caloric value (kJ/100g) = P x 77 + F x 37 + C x 16

CHAPTER 3

ANALYSIS OF LIPID COMPONENTS

Lipids are one of the major constituents in fish and shellfish that provide energy and essential nutrients. Nevertheless, over-consumption of cholesterol and saturated fatty acids is detrimental to health. Lipids determine the physical properties of fish, such as flavor, texture, mouthfeel, and appearance. Lipids prone to oxidation may lead to the formation of off-flavors and harmful products. Therefore, food analysts need to know more about the properties of lipids.

- Total lipid concentration
- Type of lipid classes
- Type of fatty acids
- Lipid oxidation products
- Chemical properties of lipids
- Physicochemical properties of lipids

The analysis of total lipids from fish by destructive and non-destructive methods is given detail in the previous chapter 2. This chapter discusses the lipid classes, individual fatty acids, lipid oxidation, and hydrolysis products of lipids.

3.1. ANALYSIS OF LIPID CLASSES

Lipids consist of complex mixtures of individual lipid classes. No single procedure gives the desired separation and so combinations of techniques to be used for the separation of different lipid classes. Low-pressure column chromatography fractionates the lipid classes initially. Further, separation of lipid achieved using silica acid or acid-washed florisil or florisil adsorbents. Column chromatography on diethyl aminoethyl (DEAE) cellulose is a valuable method for the isolation of complex lipids. Aminopropyl bonded phase cartridges isolate simple and complex lipid fractions. Thin-layer chromatography (TLC) routinely applied for the separation, identification, and quantification of lipids. A variety of solvent systems separates simple lipids on an analytical or semi-preparative scale. The most frequently used ones are hexane, diethyl ether, and acetic acid in various proportions.

3.1.1.Column chromatography method

Simple lipids are eluted in a stepwise sequence with hexane containing increasing proportions of diethyl ether, whereas complex lipids are recovered by elution with methanol.

Reagents

1. Silica gel (60-120 mesh size)
2. Chloroform
3. Hexane
4. Acetone
5. Diethyl ether

Preparation of column

Activate the silica gel for 1-2h in a hot air oven set at 100 °C. Plug the bottom of the column with cotton to prevent the leakage of silica gel. Mix 15 g of activated silica gel with a suitable solvent and pour the mixture into the column to make the bed. Use chloroform for major lipid classes, and hexane for neutral lipid separation. Insert a filter paper on the top of the bed.

Separation of major lipid fractions

Add 1 ml of the lipid solution onto the column. Then, add 100ml of chloroform over that and open the valve slowly and collect the first fraction. This fraction is "Neutral lipids". Evaporate the fraction and note down the weight. Likewise, collect the second fraction by elution with 150ml acetone: methanol (9:1). This fraction is "Glycolipids". Then, collect the third fraction by elution with 100 ml methanol. This fraction is "Phospholipids".

Separation of neutral lipids

Add 1 ml of the concentrated neutral lipid fraction back onto the column. Then, add 30 ml of hexane and elute the first fraction to obtain "Hydrocarbons". Then, elute the second fraction with 1% diethyl ether in 60 ml hexane to get "Sterol/ FAME/wax". Again, elute the third fraction with 5% diethyl ether in 50 ml hexane to obtain "Triacylglycerols/fatty acids". Then, elute the fourth fraction with 8% diethyl ether in 50 ml hexane to get "Free fatty acids". Elute the fifth fraction with 15% diethyl ether in 80 ml hexane to obtain "Diacylglycerol/cholesterol". Finally, elute the sixth fraction with 50 ml diethyl ether to get "Monoacyl glycerol/ diacyl glycerol/ others". Evaporate all the collected fractions individually and note down the weights.

3.1.2. Thin layer chromatography method

In the thin layer chromatography (TLC)method, the solute competes with the solvents for the surface sites on the adsorbent. The lipid compounds get distributed on the surface of the adsorbent silica depending on the distribution coefficients.

Reagents

1. **Developing solvents**

   Petroleum ether: diethyl ether: acetone (90/10/1) or hexane:ether:acetic acid (60:40:1)

2. Adsorbent silica gel G

3. **Standard neutral lipids:** Phospholipids, monoacyl glycerol, diacyl glycerol, free fatty acids, triacyl glycerol, cholesterol and cholesterol esters. Dissolve them in known concentration of chloroform.

4. **Detection agent:** Iodine

5. Chloroform or methanol

Preparation of plates

Clean the glass plates (20 × 20 cm), dry, and place on the plastic base plate over a plane surface. Prepare a slurry of the adsorbent by mixing 10 g of silica in 30 ml of distilled water at the ratio of 1:3 (w/v) to make a single 20 x 20 cm TLC plate. Stir the slurry thoroughly for 1-2 min and pour it into the applicator positioned on the head glass plate. Coat the slurry over the glass plates at a thickness of 0.25mm, by moving the applicator at a uniform speed from one end to the other. We can also prepare the plates by pouring 30 ml of slurry in a TLC plate and spreading uniformly using a glass rod. Allow the plates to dry at room temperature for 15-30 min and then in an oven at 100 – 120°C for 1 –2h to remove the moisture, and to activate the adsorbent on the plate. Store the dried plates in a rack in a desiccator over silica gel to prevent moisture absorption.

Sample application

Dissolve the lipid sample in known concentration of chloroform. Spot the samples at a distance of 2.5 cm away from one bottom end of the glass plate. Apply 25 ml of the sample or standard using a micropipette as small spots. Place the spots at equal distance from one end of the plate. Allow the sample to dry to do additional spotting or to have concentrated sample spot.

Chromatogram development

Pour the developing solvent into the tank to a depth of 1.5 cm and leave for at least an hour with a vacuum greased cover plate over the top of the tank to ensure that the atmosphere within the tank becomes saturated with the solvent vapor. This is knwon as equilibration. After equilibration, remove the cover plate and place the thin layer plates vertically in the tank such that the spotted end dips in the solvent. Replace the cover plate during the separation. The separation of compounds occurs as the solvent moves upward. Remove the plate from the tank once the solvent reaches the top of the plate. Dry the plates and identify the separated compounds.

Identification

Place the plates in the iodine tank in a fume hood. Observe the lipids as yellow-brown spots after about 5 min Fig. 1 gives the separation pattern of neural lipids in silica gel plate. Record the distance traveled by developing solvent and the distance traveled by each lipid fraction. Calculate the relative factor for each lipid fraction using the formula:

$$\text{Relative factor (RF)} = \frac{\text{Distance travelled by the lipid fraction}}{\text{Distance travelled by the developing solvent}}$$

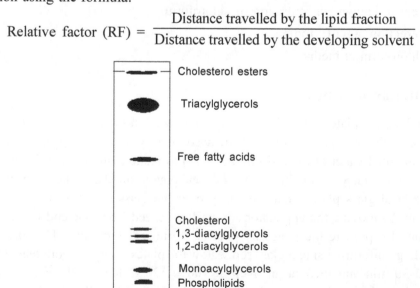

Fig. 1. Separation pattern of neutral lipid fractions in silica gel TLC plate.

Elution of samples

Mark the edges of the spots with a pencil and scrap off the lipid fractions and place them in weighing paper, which is folded and rolled to grind the clumps. Prepare the columns to elute the lipid fractions from the silica gel. For which, insert a small amount of glass wool into a pipette, label, and leave on the stand. Carefully transfer the silica gel having different lipids into the appropriate columns. Keep the glass test tubes beneath the columns and add 1 ml of chloroform into each column except for the phospholipid and monoacylglycerol columns. To the latter, add 1 ml of methanol. Shake the test tubes and place in the fume hood. Evaporate the solvent off under a stream of N_2 and store the dry lipids under N_2.

3.2. Analysis of Fatty Acids

Fatty acids are essential components of lipids. Gas-liquid chromatography (GC) is the most commonly used method for fatty acid analysis. Lipid extracts are converted into methyl ester derivatives by acid catalysis and base catalysis.

In acid catalysis, methylate free fatty acids and trans-methylate o-acyl lipids by heating them with an excess of anhydrous methanol in the presence of an acidic catalyst. Transesterify fatty acids in amide bound sphingolipids, and liberate aldehydes from plasmalogens under acidic conditions. The reagent for esterification is anhydrous hydrogen chloride in methanol. Fatty acid methyl esters (FAMEs) are formed by heating the lipids with an esterifying agent at 50°C overnight. Boron trifluoride (BF_3) in methanol is used as a transmethylation catalyst to esterify free fatty acid (FFA).

In basic catalysis, trans-esterify o-acyl lipids in anhydrous methanol with a catalyst. The reagent is sodium methoxide in anhydrous methanol.

3.2.1. Gas chromatography method

Fatty acids are complex molecules consisting of acylglycerols, cholesterol esters, waxes, and glycosphingolipids. Fatty acids are released from complex molecules by saponification or acidic hydrolysis. Lipids are hydrolyzed in an alkaline medium to extract unsaponifiable sterols, alcohols, hydrocarbons, pigments, and vitamins (AOAC, 1990). Non-reactive derivatives of fatty acid methyl esters (FAMEs) are more volatile than the free acid components. Lipids are transformed by a transesterification reaction that displaces the glycerol moiety by another alcohol (methanol, butanol, etc.) in acidic conditions (HCl or BF_3). The most common derivatives of fatty acids are the methyl esters.

Fatty acid methyl esters are obtained by heating free fatty acids with a large excess of anhydrous methanol in the presence of a catalyst, boron trifluoride, BF_3. The o-acyl glycerols (lipids) gets trans esterified very rapidly.

$$\Rightarrow \quad \xrightleftharpoons{H^+} RCOOCH_3 + R'OH$$

$$RCOOH + CH_3OH \xrightleftharpoons{H^+} RCOOH_3 + H_2O$$

Reagents

1. Toluene
2. Alcoholic NaOH, 0.5 M
3. Saturated sodium chloride, 25%
4. BF_3- Methanol
5. HPLC grade methanol
6. Fatty acid standard: Sigma FAME mixture (C4 – C24)

Extraction and methylation

Extract the total lipid from 10 g of fish tissue by Folch's method or extract the lipid classes by column chromatography or TLC, as described in 3.1.1 and 3.1.2 for direct esterification. Take 250mg of lipid fraction and dissolve in toluene in a round bottom flask. To which, add 4 ml of alcoholic NaOH and reflux for 5 – 10 min until droplets of fat disappear. Then, add 5 ml of BF_3 methanol and reflux for another 1 min. Cool the content and add 15 ml of saturated sodium chloride (25%) for purification of esters. Extract the content with 5 ml of hexane twice and remove the upper hexane layer, combine and evaporate to dryness in a rotary flask evaporator set at 55 - 60°C. Re-constitute the methyl esters in 1ml of HPLC grade hexane for injection in Gas Chromatograph.

Gas chromatograph

GC system	Perkin Elmer Gas Chromatograph
Column:	PE-225 (50% cyanopropyl phenyl 50% dimethyl siloxane copolymer) with 0.25 mm inner diameter (i.d.) and 30 m length or similar column supplied by the company

Injector temperature : 250°C

Detector : Flame ionization detector (FID)

Detector temperature : 300°C

Carrier pressure : 20 psi

Split ratio : 1:50

Programming : Gradient temperature programming

Rate	Temperature	Hold
–	70°C	1 min
8°C/min	180°C	1 min
3°C/min	210°C	30 min

Instrumental GC analysis

Equilibrate the column at 210°C for 30 min to 1 h. Inject 0.5 ml of standard fatty acid methyl ester (FAME) mix on to the GC. Separation of fatty acid methyl esters takes 45 min. Then, inject 0.5 ml of sample FAME on to the GC. Identify the fatty acids present in the sample by comparing the retention time of the fatty acids in the standard mixture. Calculate the quantity of the individual fatty acid with the help of authentic standards and express as mg of fatty acids/ 100 g of fat or as area percentage.

3.3. ANALYSIS OF LIPID OXIDATION PRODUCTS

Fish lipids are good sources of polyunsaturated n-3 fatty acids (PUFA) such as eicosapentaeonic acid (EPA, C20:5n-3) and docosahexaenoic acid (DHA, C22:6n-3). A high proportion of long-chain PUFAs makes fish lipids susceptible to oxidation and limits the shelf life of marine oils and fish products. The reaction products of lipid oxidation affect the sensory properties of fish products.

Lipid oxidation is of three types: autoxidation, photooxidation, and enzymatic oxidation. Unsaturated fatty acids, when exposed to oxygen, cause autoxidation. Hydrogen ions get extracted from the fatty acids to form free radicals. The free radicals react with oxygen to form peroxy radicals and hydroperoxides. The peroxides easily break down to alkoxyl radicals leading to the formation of a wide variety of non-radical reaction products such as aldehydes, ketones, acids, and alcohols; and also, more complex reaction products such as epoxy and polymeric compounds. Fatty acids and lipid oxidation products react with other food components such as proteins, carbohydrates, and water. Different methods are available to determine the degree of lipid oxidation, particularly the primary and secondary oxidation products.

3.3.1. Primary oxidation products

Primary oxidation products of fatty acid are peroxide value and conjugated dienes.

3.3.1.1. Peroxide value

Peroxides (R-OOH) form in the initial stages of oxidation as primary reaction products and indicate the progress of lipid oxidation. Determination of peroxide value (PV) is either by titrimetric or spectrophotometric method. Before the analysis, the lipids are extracted using suitable solvent (Folch *et al.*, 1957 or Bligh and Dyer, 1959).

3.3.1.1.1. Titration method

The titration method determines the peroxide value of a lipid. The sensitivity of this method is 0.5 meq/kg lipid. A sample size of 5 g is required if the PV is below 10 and 1 g if the PV is high. Peroxide value is expressed in milliequivalent of peroxide per kg of lipid (AOCS, 1995). Oxygen, light, and absorption of iodine by the unsaturated fatty acids are some interferences. **PV should not be above 10–20 meq/kg of fish lipid**.

The method utilizes the ability of peroxides to liberate iodine from potassium iodide. The lipid is dissolved in a suitable organic solvent and an excess of KI is added:

$$ROOH + KI_{excess} \Rightarrow ROH + KOH + I_2$$

Once the reaction is complete, the amount of ROOH that has reacted can be determined by measuring the amount of iodine formed. This is done by titration with sodium thiosulfate and a starch indicator:

$$I_2 + starch + 2Na_2S_2O_3 \text{ (blue)} \Rightarrow 2NaI + starch + Na_2S_4O_6 \text{ (colourless)}$$

The amount of sodium thiosulfate required to titrate the reaction is related to the concentration of peroxides.

Reagents

1. Glacial acetic acid
2. Chloroform

3. Potassium iodide (KI),10%

4. Anhydrous sodium sulfite

5. Starch, 1% (Prepare fresh daily)

6. Sodium thiosulfate, 0.02 N

Procedure

Weigh accurately 10±0.1g of sample and homogenize with 15 g of anhydrous sodium sulfite to remove the moisture. Extract the lipid with 50 ml of chloroform, and filter. Take 15 ml of the filtrate in a conical flask, add 15 ml of glacial acetic acid and 10 ml of 10% KI solution. Keep the mixture in the dark place for 10 min with occasional shaking. Then, add 50 ml of distilled water and 1 ml of starch solution. Titrate the liberated iodine with standard sodium thiosulfate until the disappearance of blue color. To estimate the lipid content in the chloroform extract used for titration, take 15 ml of the extract in a pre-weighed beaker and evaporate to dryness by placing in a water bath or hot plate. Perform the final drying at 100°C in a hot air oven, cool in a desiccator, and weigh again. The difference in the initial and final weight gives the lipid content. The volume of chloroform extract taken for titration and lipid estimation should be same.

Calculation

$$\text{Peroxide value (milli equi of peroxides /kg of lipid)} = \frac{T.V \times N \times 1000}{W}$$

where, T.V – titer value, ml; N – normality of sodium thiosulphate, 0.02; W – weight of the fish lipid, g; 1000- kg conversion

3.3.1.1.2. Spectrophotometric method

Ferric thiocyanate method is a spectroscopic method used to determine the PV of the lipids (Mihaljevic *et al.*, 1996). This method is more sensitive and requires less sample size (10 mg). This method is also known as the International Dairy Federation (IDF) method (Sato *et al.*, 1991).

Peroxides present in the lipid oxidize ferrous ions to ferric ions, which react with ammonium thiocyanate to form ferric thiocyanate, which gives a red complex having an absorption maximum at 500 nm.

Reagents

1. Chloroform: methanol (2:1) and (7:3)

2. Barium chloride dehydrate

3. Ferrous sulfate

4. Hydrochloric acid, 10 N

5. Ammonium thiocyanate, 30 %

6. **Ferric chloride ($FeCl_3$) solution (0.02 M):**

   Dissolve 0.4 g of barium chloride dihydrate in 50 ml of distilled water. Dissolve 0.5 g ferrous sulfate in 50 ml of distilled water. Slowly add barium chloride solution with constant stirring to the ferrous sulfate solution. Finally, add 2 ml of 10 N HCl to the above solution. Centrifuge at 3000 rpm for 5 min. Pipette out clean iron (II) solution, leaving behind the barium sulfate precipitate. Store iron (II) solution in brown bottle and keep in the dark.

7. **Fe^{3+} standard solutions**

   Stock (1 mg/ml Fe^{3+}): Dissolve 2.47 g of ferric chloride in 500 ml of 10 N HCl. Store in an amber-glass bottle at 4°C.

   Intermediate Standard (0.1 mg Fe/ml): Add 1 ml of stock to 9 ml $CHCl_3$: MeOH (2:1). Mix well. (Prepare fresh daily)

   Working Standard 1 (10 µg Fe /ml): Mix together 3 ml of intermediate standard and 27 ml $CHCl_3$: MeOH (2:1) (Prepare fresh daily)

   Working Standard 2 (1µgFe /ml): Mix together 3 ml of working standard 1 with 9 ml of $CHCl_3$: MeOH (2:1) (Prepare fresh daily)

   Working Standard 3 (0.5 µg Fe/ml) : Mix together 5 ml of working standard 2 with 5 ml of $CHCl_3$: MeOH (2:1) (Prepare fresh daily)

Procedure

Weigh accurately 0.001 to 0.03g of lipid or oil sample in a test tube and add 9.8 ml of chloroform: methanol (7:3 v/v) mixture. Then, add 0.1 ml of ferric chloride solution and 0.1 ml of 30% ammonium thiocyanate solution. Mix the content well and place it in dark for 10 min. Similarly, analyze the standard solution to construct a Fe^{3+} standard curve, as given in the Table. Measure the absorbance (O.D) at 500 nm in a spectrophotometer. Plot the absorbance value on the X-axis and standard concentration on the Y-axis. Determine the regression equation for the curve using any standard program, to know the slope (m) and y-intercept.

Standard Fe^{3+} Conc. (μM)	10 μg Fe/ml std (μl)	1 μg Fe/ml std (μl)	0.5 μg Fe/ml std (μl)	Chloroform: Methanol 2:1 (μl)	30% SCN (μl) *	Total (μl)
1.013	0	0	300	2337.5	12.5	2650
1.689	0	0	500	2137.5	12.5	2650
2.365	0	0	700	1937.5	12.5	2650
3.378	0	0	1000	1637.5	12.5	2650
5.067	0	0	1500	1137.5	12.5	2650
7.770	0	0	2300	337.5	12.5	2650
13.51	0	2000	0	637.5	12.5	2650
40.54	600	0	0	2037.5	12.5	2650
54.053	800	0	0	1837.5	12.5	2650
67.566	1000	0	0	1637.5	12.5	2650
135.133	2000	0	0	637.5	12.5	2650
168.916	2500	0	0	137.5	12.5	2650

*Note: Add 30% SCN (ammonium thiocyanate) last.

Calculation

$$\text{Peroxide value} = \frac{As\text{-}Ab \times m}{55.8 \times 2 \times W}$$

where, As – absorbance of sample; Ab – absorbance of blank; m – slope of calibration curve; W – Weight of the lipid, g

There are many problems with the use of peroxide value as an indication of lipid oxidation. First, the peroxides are primary products that break down in the latter stages of lipid oxidation. Thus, a low value of PV may represent either the initial or final stages of oxidation. Second is that the results of the procedure are highly sensitive to the conditions used to experiment. This technique is an example of a measurement of the increase in the concentration of primary reaction products.

3.3.1.2. Conjugated dienes

Oxidation of PUFA leads to the formation of conjugated diene hydroperoxides. The structure consists of a fatty acid chain with alternating simple and double bonds and has an absorption maximum at 230-235 nm. Conjugated dienes estimated in the oil sample by diluting in methanol, isooctane, or hexane. In tissue extracts, extraction, and separation techniques to be employed for their estimation (AOCS, 1995).

3.3.2. Secondary oxidation products

Peroxides are unstable molecules that get rapidly transformed into secondary oxidation products such as aldehydes, ketones, and small organic acids. Two methods are employed to determine the presence of aldehydes: thiobarbituric acid-reactive substances (TBARS) and anisidine value (AnV).

3.3.2.1. Anisidine value

Anisidine value (AnV) is the common method for the determination of secondary oxidation products (IUPAC, 1987). It determines the number of aldehydes (mainly 2-alkenals and 2,4-dienals) generated during the decomposition of hydroperoxides. It is more sensitive to unsaturated (core) aldehydes than to saturated aldehydes. The p-AnV gives a reliable indication of oxidative rancidity in fats, oils, and fatty foods and employed in Europe along with the Totox number.

Unsaturated aldehydes in the lipids react with p-anisidine dissolved in acetic acid and gives yellow colored product having an absorption maximum at 350 nm.

Reagents

1. Isooctane or n-hexane
2. **p-anisidine reagent:** Dissolve 0.25 g of anisidine in 10 ml of glacial acetic acid and protect it from light

Procedure

Take about 1±0.1 g of lipid in 25 ml of isooctane or n-hexane to make the test solution A. Take 5 ml of the test solution A in a test tube and add 1.0 ml of p-anisidine reagent to make the test solution B. Then, add 5 ml of isooctane or n-hexane in a test tube and add 1.0 ml of p-anisidine reagent to serve as a blank. Leave the test tubes at room temperature for 10 min. Read the absorbance (O.D) at 350nm using a spectrophotometer.

Calculation

$$\text{Anisidine value} = \frac{25 \times (1.2\ A_1 - A_2)}{W}$$

where, A_1- absorbance of test solution A; A_2- absorbance of test solution A; W – weight of the lipid, g

3.3.2.2. TOTOX value

TOTOX value is used in commercial laboratories to determine fat oxidation products. A combination of PV and AnV gives the Totox value (2*PV + AnV).

3.3.2.3. TBARS value

Thiobarbituric acid reactive substances (TBARS) value determines the secondary oxidation products comprising aldehydes, ketones, short-chain fatty acids, etc. A micro method is available for the determination of malonaldehydes following steam distillation (Ke and Woyewada, 1979). **The TBARS value should not be above 1 – 2 m mole malonaldehyde per gram lipid or above 10 -20 m mole malonaldehyde per kg of lipid in seafood.**

Direct distillation of sample yields carbonyls, the oxidation products of PUFA. Thiobarbituric acid (TBA) reacts particularly with an aldehyde, one of the carbonyl compounds. Condensation of two moles of TBA and one mole of malonaldehyde under acidic conditions forms a red color complex having an absorption maximum at 538 nm. TBARS represents different secondary oxidation products.

Reagents

1. Glacial acetic acid

2. Hydrochloric acid, 4 N

3. **TBA reagent:** Add 1.44 g of 2- thiobarbituric acid and 50 ml distilled water in a 500 ml volumetric flask. Add 50 ml of glacial acetic acid and stir well to dissolve TBA completely and make up to the mark with glacial acetic acid

4. **Standard TEP solution**

   Stock (1 x 10^{-2} M): Weigh 0.22 g of 1,1,3,3, - tetra ethanol propane (TEP) into 100 ml volumetric flask and dilute to volume with distilled water.

   Working standard,I (1 x 10^{-4}M): Pipette 1 ml of this solution into 100 ml volumetric flask and dilute to volume to produce 1 x 10^{-4} M solution.

   *Working standard II (1 x 10^{-5}M):*Pipette 10 ml of working I solution and dilute to 100 ml with distilled water.

5. Propyl gallate and EDTA.

Procedure

Weigh about 10 g of the sample and homogenize with 35 ml of distilled water. Transfer the homogenate to a 500 ml round bottom flask and add 100 mg of propyl gallate and EDTA. Then, add 50 ml of distilled water and 25 ml of 4 N HCl to the homogenate. Distill the sample and collect 50 ml of distillate. Take 5 ml of the sample distillate in a screw cap test tube. Take 5 ml of distilled water to serve as a blank. Take 0.4, 0.8, 1.2, 1.6, and 2.0 ml of working standard II into a series of screw cap test tubes and adjust the volume to 5 ml with distilled water. The concentration of TEP in 10 ml of each standard corresponds to 4, 8, 12, 16, and 20 x 10^{-7} M per liter, respectively. Add 5 ml of TBA reagent to all the tubes and cover them, mix, and heat in a boiling bath (100°C) for 45 min and cool. Read the absorbance (O.D) at 538 nm in a spectrophotometer within 30 min. Draw a standard graph by plotting the absorbance on the Y-axis and the concentration of TEP on the X-axis. Calculate the concentration of TEP present in the sample from the graph.

Calculation

If the volume of aliquot is 5ml

TBA (m mole/kg of fish) = Conc of TEP in sample $\times 10^7$

or

If the volume of aliquot use is different from 5ml

$$\text{TBA (mmole/kg of fish)} = \frac{5ml}{\text{Aliquot size}} \times \text{Conc. of TEP in sample} \times 10^7$$

where, 10^7 - μ mole and kg, conversion.

3.4. ANALYSIS OF OTHER IMPORTANT LIPID OXIDATION PRODUCTS

There are several other methods for the determination of lipid oxidation- electronic nose array system, fluorescence method, stability, and instrument method. Electronic noses or gas sensory array system collects and identifies headspace volatiles. The fluorescence method identifies fluorescent products formed due to the interaction of lipids with proteins.

3.4.1. Fluorescence method

Lipid oxidation products interact with other food components such as amino acids, peptides, proteins, nucleic acids, DNA, phospholipids, etc. to form fluorescent compounds. Fluorescent compounds form as the result of oxidation of phospholipids or from oxidized fatty acids in the presence of phospholipids. The fluorescent method is 10-100 times more sensitive than the TBARS method for the detection of malonaldehydes. This method applied to assess lipid oxidation during fish processing traditionally (Aubourg and Medina, 1999).

The fluorescent compounds formed due to various interactions of lipids have different excitation and emission maxima. The fluorescence shift is the most effective index to determine the changes in fish quality than the other commonly used methods. The fluorescence shift is calculated by dividing the fluorescence intensities of the compounds with that of the fluorescence intensity of quinine sulfate. The fluorescence method can distinguish between different oxidation products formed in fish.

Reagents

1. Chloroform

2. Methanol

3. Quinine sulfate: Prepare 1 µg/ml solution in 0.05 M H_2SO_4

4. Sulfuric acid, H_2SO_4, 0.05M

Procedure

Take about 10g of wet muscle tissue and extract with 30 ml of chloroform: methanol: water (2:2:1.8) and collect the aqueous phase for fluorescence determination. Measure the fluorescent intensity in the extract using a spectrofluorometer at 393/463 nm and 327/415 nm of excitation emission wavelengths. Then, measure the fluorescent intensity of the quinine sulfate solution at the same wavelengths. Calculate the relative fluorescence and fluorescence ratio. Fluorescence ratio (FR) is the ratio between the two-relative fluorescence (RF) values.

Calculation

$$\text{Relative fluorescence (RF)} = \frac{F}{Fst}$$

where, F- fluorescence measured at each excitation/emission maximum;

Fst- fluorescence intensity of a quinine sulfate solution at the same wavelengths

$$\text{Fluorescence ratio (FR)} = \frac{\text{RF393/463 nm}}{\text{RF327/415 nm}}$$

3.4.2. Stability methods

There are methods to evaluate oxidation based on accelerated oxidation: oil stability index (OSI) method, rancimat test, oxidograph, and active oxygen method (AOM). In the rancimat and OSI methods, the oil heated at 80°C or more form low molecular weight acids when air bubbles pass through and get collected in vessels containing distilled water. The measure of the change in conductivity gives the point where it changes the most is the "induction time". The AOM method measures the time taken to reach a particular peroxide value (PV). The oxidograph instrument measures the induction time based on the decline in pressure caused by the absorption of oxygen in a closed vessel.

3.4.3. Instrumental methods

The methods used for the determination of oxidation parameters in oils and food: Near-infrared spectroscopy (NIR), Fourier transform near-infrared spectroscopy (FT-NIR), and FT-IR spectroscopy method. Electron spin resonance (ESR) spectroscopy assesses the free radicals to detect the early stages of lipid oxidation. Gas chromatography-mass spectrometry (GC-MS) method determines a wide range of volatile secondary lipid oxidation products, while liquid chromatography-mass spectrometry (LC-MS) method can determine non-volatile products. ^{1}H-NMR spectra determine specific lipid oxidation products such as different hydroperoxides, aldehydes, and cyclic compounds.

3.5. ANALYSIS OF CHEMICAL PROPERTIES OF LIPIDS

A number of chemical methods are developed to provide information about the type of lipids present in edible fats and oils. These techniques give information about the average chemical properties of the lipid components. average molecular weight, degree of unsaturation or amount of acids. Nevertheless, these methods are simple to perform and do not require expensive apparatus.

3.5.1. Iodine Value

The iodine value (IV) gives a measure of the average degree of unsaturation of a lipid. The higher the iodine value, the greater the number of C=C double bonds. Iodine value defined as the grams of iodine absorbed per 100g of lipid. The "Wijs method" is the most common method used to determine the iodine value of lipids.

In this method, iodine chloride added in excess to the lipids dissolved in a suitable organic solvent. Some of the iodine and chlorine (ICl) reacts with the double bonds in the unsaturated lipids, while the rest remains:

$$R\text{-}CH\text{=}CH\text{-}R + ICl_{excess} \Rightarrow R\text{-}CHI\text{-}CHCl\text{-}R + ICl_{remaining}$$

The amount of $ICl_{reacted}$ is determined by measuring the amount of ICl remaining after the completion of the reaction ($ICl_{reacted} = ICl_{excess} - ICl_{remaining}$). The amount of $ICl_{remaining}$ is determined by adding excess potassium iodide to the solution to liberate iodine, and then titrating with a sodium thiosulfate ($Na_2S_2O_3$) solution in the presence of starch to determine the concentration of iodine released

$$ICl_{remaining} + 2KI \Rightarrow KCl + KI + I_2$$

$$I_2 + starch + 2Na_2S_2O_3 \text{ (blue)} \Rightarrow 2NaI + starch + Na_2S_4O_6 \text{ (colourless)}$$

Iodine has a reddish-brown color, but not intense enough as a good indication of the end-point of the reaction. Starch is as an indicator that forms a molecular complex with the iodine and produces a deep blue color. The addition of starch to the solution containing iodine turns it to dark blue initially. Perform the titration with sodium thiosulfate solution of known molarity. If any I_2 remains in the solution, it stays blue, and once all of the I_2 gets converted to I., it turns colorless. Thus, a change in solution appearance from blue to colorless gives the end-point of the titration. The concentration of C=C in the original sample determined based on the amount of sodium thiosulfate needed to complete the titration.

The higher the degree of unsaturation, the higher is the iodine absorbed, and the higher is the iodine value. The iodine value is a measure of the average degree of unsaturation of oils. Hydrogenation and oxidation processes cause changes in the degree of unsaturation and alter the iodine value.

3.5.2. Saponification Number

The saponification number gives the average molecular weight of the triacylglycerols in a sample. Saponification is the process of breaking down a neutral fat into glycerol and fatty acids by treatment with alkali.

Triacylglycerol + 3 KOH ⇨ Glycerol + 3 Fatty acid salts of potassium

The saponification number defined as the mg of KOH required to saponify one gram of lipid. Extract the lipid first and dissolve in an ethanol solution that contains a known excess of KOH. Heat this solution until the reaction is complete. Determine the unreacted KOH by the addition of an indicator followed by titration with HCl. Calculate the saponification number from the weight of the sample, and the amount of KOH reacted. The smaller is the saponification number, the larger, is the molecular weight of the triacylglycerols.

3.5.3. Acid value or FFA value

The acid value is a measure of the number of free acids present in a given amount of lipid. Extract the lipids from the food sample and dissolve in an ethanol solution containing an indicator. Titrate the solution with standard alkali (KOH or NaOH) until a pinkish color appears. The acid value is the mg of KOH required to neutralize the fatty acids present in 1g of lipid. The acid value becomes overestimated, if other acid components are present in the system, *e.g.* amino acids or acid phosphates. The acid value also gives a good measure of the breakdown of the triacylglycerols into free fatty acids, which has an adverse effect on the quality of many lipids.

Lipid undergoes hydrolysis by the action of lipases present in the fish. The liberation of free fatty acid (FFA) and glycerol leads to hydrolytic rancidity in fish. FFA and their oxidation products affect fish muscle texture and functionality, as they interact with myofibrillar proteins and promote protein aggregation. A gradual increase in FFA formation occurs during the frozen storage of marine fatty and lean fish; and also freshwater fish, indicating the lipid quality.

Free fatty acids liberated upon hydrolysis of fat are estimated by titrating them against standard alkali in the presence of meta-cresol purple indicator. The free fatty acid content is expressed as oleic acid equivalents.

Reagents

1. NaOH, 0.05N
2. Metacresol purple, 0.5% aqueous
3. Chloroform, methanol and isopropanol (2:1:2)
4. Anhydrous sodium sulfite

Procedure

Weigh 10±1g of tissue, homogenize, and transfer to a conical flask. Add 15 g of anhydrous sodium sulfate to remove moisture. Add 30 ml of chloroform: methanol: isopropanol mixture, mix well and filter after 15 min to get the extract. To 10 ml of the extract taken in a flask, add 3 drops of 0.5% metacresol purple indicator and titrate the content with NaOH solution until pink color endpoint appears. In another pre-weighed beaker, take 10ml of extract, and evaporate the lipid by placing over a water bath. Calculate the lipid present in the extract based on the difference in the initial and final weights. If the sample is an oil, use 1g of sample for extraction.

Calculation

$$\text{Free fatty acid (\% as oleic acid)} = \frac{T.V \times N \times 2.82 \times D.F}{W}$$

where, T.V – titer value, ml; N - normality of NaOH (0.05N); D.F – dilution factor; W - weight of fish, g; 2.82 – amount of NaOH required to neutralize FFA present in 1g of lipid

3.6. ANALYSIS OF PHYSICO-CHEMICAL PROPERTIES OF LIPIDS

Besides nutritional importance, lipids have characteristic physicochemical properties such as mouthfeel, flavor, texture, and appearance. They are heat transfer agents in the preparation of other foods, e.g., for frying. Several analytical techniques are available to characterize the physicochemical properties of lipids.

3.6.1. Solid Fat Content

Solid fat content (SFC) of a lipid influences its sensory and physical properties, such as spreadability, firmness, mouthfeel, processing, and stability. Food manufacturers measure the variation of SFC with temperature when characterizing lipids used in certain foods, e.g., margarine and butter. The solid fat content is defined as the percentage of the total lipid, i.e. solid at a particular temperature.

$$SFC = \frac{M_{solid}}{M_{total}} \times 100$$

where M_{solid} is the mass of the lipid that is solid;

M_{total} is the total mass of the lipid in the food

There are many methods employed to measure the temperature dependence of solid fat content. The density of solid fat is higher than that of liquid oil. When the density increases, the fat crystallizes and when it decreases, the fat melts. The solid fat content - temperature profile is determined by measuring the densities over a wide temperature range. The density measured either by density bottles or dilatometer.

$$SFC = \frac{(\rho - \rho_L)}{(\rho_S - \rho_L)} \times 100$$

where, ρ is the density of the lipid at a particular temperature, ρ_L and ρ_S are the densities of the lipid, if it were completely liquid or completely solid at the same temperature.

Nuclear magnetic resonance (NMR) method had replaced density measurements recently, as it is quicker and simpler. In NMR, signal induced by placing the sample into the instrument and applying a radiofrequency pulse. The decay rate of the signal depends on whether the lipid is solid or liquid. The signal from the solid lipid decays rapidly than the liquid oil, and therefore, it is possible to distinguish between these two samples. Techniques based on differential scanning calorimetry are commonly used to monitor changes in SFC. These techniques measure the heat evolved or absorbed by a lipid when it crystallizes or melts. By making these measurements over a range of temperatures, the melting point, the total amount of lipids involved in the transition, and the SFC-temperature profile are determined.

3.6.2. Melting point

In many situations, it is not necessary to know the SFC over the whole temperature range, instead only information about the temperature at which melting starts or ends is required. A pure triacylglycerol has a single melting point that occurs at a specific temperature. Nevertheless, food lipids contain a wide variety of different triacylglycerols, each with their unique melting point, and so they melt over a wide range of temperatures. Thus, the "melting point" of a food lipid can be defined in many different ways, each corresponding to a different amount of solid fat remaining. Some of the most commonly used "melting points" are:

- *Clear point:* Place a small amount of lipid in a capillary tube and heat at a controlled rate. The temperature at which the lipid completely melts and becomes transparent is the "clear point".

- *Slip point.* Place a small amount of lipid in a capillary tube and heat at a controlled rate. The temperature at which the lipid starts to move downwards due to its weight is the "slip point".

- *Wiley melting point.* Suspend a disc of lipid in an alcohol-water mixture of similar density and heat at a controlled rate. The temperature at which the disc changes shape to a sphere is the "Wiley melting point".

3.6.3. Cloud point

Cloud point gives a measure of the temperature at which crystallization begins in the liquid oil. Heat a lipid sample to a temperature where all the crystals melt (*e.g.,* 130°C). Cool the sample at a controlled rate and determine the temperature at which the liquid goes cloudy. The "cloud point" is the temperature where crystals begin to form and scatter light. The oil that does not crystallize when stored at 0°C for prolonged period has practical importance. It is a simple test to determine the ability of lipids to withstand cold temperatures without forming crystals.

3.6.4. Smoke, Flash and Fire Points

These tests give a measure of the effect of heating on the physicochemical properties of lipids. They help to select the lipids for usage at high temperatures, *e.g.,* during baking or frying. They reflect the amount of volatile organic material in oils and fats, such as free fatty acids.

- *Smoke Point:* The temperature at which the sample begins to smoke when tested under specified conditions. Pour a lipid into a metal container and heat at a controlled rate in an oven. The smoke point is the temperature at which a thin continuous stream of bluish smoke observed first.

- *Flash point:* The temperature at which a flash appears at any point on the surface of the sample, due to the ignition of volatile gaseous products. Pour the lipid a metal container and heat at a controlled rate, with a flame passing over the surface of the sample at regular intervals.

- *Fire point:* The temperature at which evolution of volatiles due to the thermal decomposition of the lipids proceed so quickly that continuous combustion occurs (i.e. a fire).

3.6.5. Rheology

The rheology of lipids is important in food applications. Rheology is the science concerned with the deformation and flow of matter. Most rheological tests involve applying a force to material and measuring its flow or change in shape. Many of the textural properties that people perceive when they consume foods are largely rheological in nature, *e.g.,* creaminess, juiciness, smoothness, brittleness, tenderness, hardness, etc. The stability and appearance of foods depend on the rheological characteristics of their components. The flow of foods through pipes or the ease at which they pack into containers determine their rheology. Liquid oils characterized in terms of their flow properties (viscosity), while viscoelastic or plastic "solids" in terms of both their elastic modulus and flow properties. A wide variety of experimental techniques are available to characterize the rheological properties of food materials.

An important rheological characteristic of lipids is "plasticity" that determines their "spreadability". The plasticity of a lipid is that fat crystals can form a three-dimensional network that gives the product some solid-like characteristics. The product behaves like a solid without any disruption to crystal network, with an elastic modulus below certain stress (known as "yield stress"). But, above this stress, the product flows like a liquid due to continued disruption of the crystal network. Rheological techniques are therefore needed to measure the change in deformation of a lipid when stresses are applied.

CHAPTER 4

ANALYSIS OF PROTEIN COMPONENTS

Proteins are polymers of amino acids. There are twenty amino acids present in proteins. Proteins differ from each other according to the type, number, and sequence of amino acids that make up the polypeptide backbone. Due to this, they differ in molecular structures, nutritional attributes, and physicochemical properties. Proteins are the nutritional constituent of foods. Food protein contains essential amino-acids such as lysine, tryptophan, methionine, leucine, isoleucine, and valine. Essential amino acids are amino acids that cannot be synthesized by the body but vital for human health. Proteins form the major structural components of many natural foods and determine their overall texture, *e.g.,* tenderness of meat or fish products. Some isolated proteins are ingredients in food because of unique functional properties, *i.e.,* desirable appearance, texture, or stability. Proteins are also gelling agents, emulsifiers, foaming agents, and thickeners. Many food proteins serve as enzymes for enhancing the rate of biochemical reactions. These reactions can either do a favorable or detrimental effect on the overall properties of foods.

Fish provides about 14% of the world's need for animal protein and 5% of the total protein requirements. Amino acid composition and digestibility of fish proteins are good. Fish is an excellent source of high-quality protein, because of the presence of essential amino acids such as lysine and methionine. Fish proteins possess functional properties such as water holding capacity, gelling, emulsification, and texture. The methods for the analysis of total protein concentration, type of the proteins, amino acids, molecular structure, and functional properties of the protein in food are therefore important. Chapter 2 dealt with the methods for the determination of total protein concentration, and this chapter describe the remaining methods of analysis.

4.1. ANALYSIS OF PROTEIN CLASSES

Food analysts are interested in the type of proteins present in a food because each protein has unique nutritional and physicochemical properties. Separation and isolation of individual proteins from a complex mixture determines the protein type, which can then be subjected to identification and characterization. Separation of proteins is based on differences in their physicochemical properties, such as size, charge, adsorption characteristics, solubility, and heat stability. Large-scale methods available for the isolation of large quantities of crude proteins, whereas small-scale methods for proteins that are expensive or available in small quantities. One of the factors to be considered during the separation process is the possibility to preserve the native three-dimensional structures of the protein molecules.

Separation of proteins is done based on the differences in their solubility in solutions. Fish muscle proteins are classified into three groups based on the solubility: sarcoplasmic, myofibrillar, and stroma proteins. Sarcoplasmic proteins, also known as water-soluble proteins, consist mainly of enzymes and are extracted using water or buffers of low ionic strength, 50mM phosphate buffer. Myofibrillar proteins, also known as salt soluble proteins, consist of myosin, actin, tropomyosin, troponin and are extracted with buffers having an ionic strength >0.3M. Stroma proteins, also known as connective tissue proteins, consist of collagen and are extracted using alkali or acid. Changes in solubility measure the changes in protein structure caused due to denaturation of protein during the storage and processing of food. The separation of different classes of protein by fractional extraction processes is described.

Reagents

1. Phosphate buffer (pH 7.0), 50 mM
2. Potassium chloride, 0.5M
3. NaOH, 0.1 M
4. Acetic acid, 0.5M

Procedure

Weigh, 5±0.5 g of fish sample, and homogenize with 80 ml of 50mM phosphate buffer for 20 sec. Centrifuge the homogenate at 5000 xg for 20 min, and decant the supernatant, which constitutes the water-soluble protein. To the precipitate, add 80 ml of 50mM phosphate buffer containing 0.5M potassium chloride and centrifuge the homogenate at 5000 xg for 20 min. The supernatant constitutes the salt soluble protein. To the precipitate, add 10 ml of 0.1 M NaOH and centrifuge the homogenate at 5000 xg for 20 min. Repeat this extraction step five times and pool all the supernatant. This fraction constitutes the alkali-soluble protein. To the precipitate, add 10 ml of 0.5 M acetic acid, leave at room temperature for 2 days, and then centrifuge the homogenate at 5000 xg for 20 min. The supernatant constitutes acid-soluble collagen fraction. Determine the absorbance of each protein fraction in the far UV range at 205 nm in a spectrophotometer or determine the protein content by Kjeldahl method. Calculate the protein concentration, based on Beer-Lambert law, if determined by spectrophotometer as A= εcl, wherein c is the molar concentration of proteins, l is the path length in cm and ε is the molar absorption coefficient in mol cm^{-1}. Carry out all the extractions at <4°C.

$$A = \varepsilon cl$$

where, c - the molar concentration of protein; ε - molar absorption coefficient mol 1 cm^{-1}; l = path length 1 cm

4.2. SEPARATION OF PROTEINS BASED ON SOLUBILITY

Separation of proteins is done based on the differences in their solubility in aqueous solutions. The solubility of a protein depends on the amino acid sequence. The amino acid composition determines the size, shape, hydrophobicity, and electrical charge of the proteins. Proteins can be selectively precipitated or solubilized by altering the pH, ionic strength, dielectric constant, or temperature of a solution. These methods are the simplest for use when large quantities of samples are involved because they are relatively quick, inexpensive, and not influenced by other food components. Separation of proteins becomes the first step in any separation procedure because the majority of the contaminating materials get easily removed.

Prior knowledge of the effects of environmental conditions on protein structure and interactions is extremely useful when selecting the most appropriate separation technique. First is to determine the most suitable conditions for the isolation of a particular protein from a mixture of proteins (*e.g.,* pH, ionic strength, solvent, temperature, etc.), and second is to choose an ideal condition that will not adversely affect the molecular structure of the proteins.

4.2.1. Salting out process

Proteins precipitate from aqueous solutions when the salt concentration exceeds a critical level leading to the salting-out process, wherein all the water becomes "bound" to the salts, and not available to hydrate the proteins. Ammonium sulfate [$(NH_4)_2SO_4$] is commonly used because it has a high water-solubility, although other neutral salts may also be used, e.g., NaCl or KCl.

Generally, a two-step procedure maximizes the separation efficiency. In the first step, the salt used is at a concentration just below that necessary to precipitate out the protein of interest, i.e., 30% of ammonium sulfate solution. Centrifugation of the solution removes any proteins that are less soluble than the protein of interest. The salt concentration used next is to a point just above that necessary to precipitate out the protein, i.e., 40%, 50%, and 60% of ammonium sulfate solution. This process precipitates out the protein of interest and leaves more soluble proteins in solution. The main disadvantage of this method is the removal of the large concentrations of salt that contaminate the protein solution before re-solubilization of protein. The removal of salts is either by dialysis or ultrafiltration.

4.2.2. Isoelectric precipitation process

The isoelectric point (pI) of a protein is the pH where the net charge on the protein is zero. Proteins tend to aggregate and precipitate at their pI because there is no electrostatic repulsion. Proteins have different isoelectric points because of their different amino acid sequences (*i.e.,* relative numbers of anionic and cationic groups), and thus can be separated by adjusting the pH of a solution. If the pH adjusted to the pI of a particular protein, it precipitates leaving the other proteins in solution.

4.2.3. Solvent fractionation process

The solubility of a protein depends on the dielectric constant of the solution that surrounds it, as this alters the magnitude of the electrostatic interactions between charged groups. The dielectric constant of a solution decreases with the increase in the magnitude of the electrostatic interactions between charged species. This alteration tends to decrease the solubility of proteins in solution as they are less ionized. Therefore the electrostatic repulsion between them is not sufficient to prevent them from aggregation. The dielectric constant of aqueous solution lowered by the addition of water-soluble organic solvents, such as ethanol or acetone. The amount of organic solvent required to cause precipitation depends on the protein. The quantity of organic solvent required to precipitate a protein varies from about 5 to 60%. Solvent fractionation performed at 0°C or below to prevent protein denaturation. When the temperature increases, protein denaturation occurs as the organic solvents mix with water.

4.2.4. Separation of contaminating proteins

Many proteins denature and precipitate from solution when heated above a certain temperature or adjusted a solution to highly acidic or basic pH. Proteins stable at high temperatures or extremes of pH are separated based on the principle that contaminating proteins precipitate while the stable protein of interest remains in solution.

4.3. SEPARATION OF PROTEINS BASED ON ADSORPTION

Separation of compounds in adsorption chromatography is by the selective adsorption and desorption at a solid matrix. The protein mixture passes through the solid matrix contained within a column. Separation achieved based on the different affinities of proteins for the solid matrix. Affinity and ion-exchange chromatography are the two types of adsorption chromatography used for the separation of proteins. Separation achieved either through an open column or high-pressure liquid chromatography.

4.3.1. Ion Exchange Chromatography

Ion exchange chromatography relies on the reversible adsorption-desorption of ions in a protein solution to a charged solid matrix or polymer network. A positively charged matrix is an anion-exchanger because it binds negatively charged ions, anions. A negatively charged matrix is a cation-exchanger because it binds positively charged ions, cations. The pH and ionic strength of the buffer conditions adjusted to favor the maximum binding of the protein of interest to the ion-exchange column. The contaminating proteins bind less strongly and therefore pass more rapidly through the column. The protein of interest gets eluted with another buffer solution that favors the desorption of protein (*e.g.,* different pH or ionic strength).

4.3.2. Affinity Chromatography

Affinity chromatography uses a stationary phase that consists of a ligand covalently bound to a solid support. The ligand is a molecule that has a highly specific and reversible affinity for a particular protein. The sample to be analyzed passes through the column, and the protein of interest binds to the ligand, whereas the contaminating proteins pass directly through the column. The protein of interest is eluted with a buffer solution that favors its desorption from the column. This technique is the most efficient means of separating an individual protein from a mixture of proteins, but it is the most expensive because of the need to have a column with specific ligand bound to them.

Laboratory uses both the ion-exchange and affinity chromatography to separate proteins or peptides. They are used less commonly for commercial separations because they are not suitable for rapidly separating large volumes and are relatively expensive.

4.4. SEPARATION OF PROTEINS BASED ON DIFFERENCES

Proteins can be separated based on their molecular size. The molecular weights of proteins vary from 10,000 to 1,000,000 daltons (10 to 1000 KDa). In practice, the separation depends on the Stokes radius of a protein, rather than directly on its molecular weight. The Stokes radius is the average radius that a protein has in solution that depends on its three-dimensional molecular structure. For proteins with the same molecular weight, the Stokes radius increases in the following order: compact globular protein < flexible random-coil < rod-like protein.

4.4.1. Dialysis process

Dialysis separates molecules in solution by a semi-permeable membrane, which permits smaller molecules to pass through it but prevents larger ones. Place a protein solution in dialysis bag, seal, and place it into a large volume of water or buffer with constant stirring. Low molecular weight solutes flow through the bag, but the large molecular weight protein molecules remain in the bag. Dialysis is a slow process and takes about 12 h to be complete. Therefore, it is most frequently used in the laboratory. It removes salt from protein solution after the salting-out process, and changes the buffers.

4.4.2. Ultrafiltration process

Place a solution of protein in a cell containing a semi-permeable membrane, and apply pressure. Smaller molecules pass through the membrane, whereas the larger ones remain in the solution. The separation principle is similar to dialysis, but pressure is applied to make the separation quick. Semi-permeable membranes with molecular cut-off ranging from 500 to 300,000 daltons (0.5 to 300 KDa) are available. The portion of the solution retained by the cell is the retentate, while the one passes through the membrane forms the ultrafiltrate. Ultrafiltration concentrates a protein solution, removes salts, exchanges buffers, or fractionates proteins based on their size. It can be used in the laboratory and on a commercial scale.

4.4.3. Gel filtration chromatography

Native gel filtration chromatography is also known as size exclusion *chromatography*. The proteins are separated based on their size and shape by this technique. There are two types of gel filtration methods based on applied pressure: low and medium as well as high-pressure chromatography. In low and medium pressure chromatography, supporting beads are open, cross-linked 3-dimensional polymers such as agarose, dextran, cellulose, polyacrylamide, and other combinations. In high-pressure chromatography, support media are macroporous silica, porous glass, or inorganic-organic composites. The separation of proteins using AKTA-Prime column chromatography is described below.

In gel filtration chromatography, small proteins enter the pores in the beads, spend more time inside beads and emerge from the column last, while larger ones cannot, and hence moves down the column first.

The coefficient K_{av} gives the proportion of pores accessible to a particular protein.

$$K_{av} = \frac{(V_e - V_o)}{(V_t - V_o)}$$

where V_e is the elution volume of the molecule, V_o is the void volume of the column and V_t is the total volume of the column.

A standard curve can be drawn with K_{av} values of standard proteins of known molecular weights and the molecular weight of the unknown protein can be determined.

Reagents

1. Ethanol, 20%
2. Dithiothreitol, 1mM
3. Blue dextran - Dissolve 4 mg in 2 ml of equilibration buffer.
4. Sephadex-G100 or Superdex-200
5. Low calcium buffer (pH 7.5) – Equilibration buffer (A)

    Mix 20 ml of 1M Tris, 7.455 g of KCl, 6.7 ml of 0.3M ethylene glycol tetraacetic acid (EGTA) and 200 µl of 4.9 M $MgCl_2$ in 1000 ml of distilled water. Add 1 mM dithiothreitol immediately before use.
6. High calcium buffer (pH 7.5) –Buffer (B)

    Mix 20 ml of 1M Tris, 7.455 g of KCl and 2 ml of 1M $CaCl_2$ in 1000 ml of distilled water. Add 1 mM dithiothreitol immediately before use.
7. Protein standards (1000 to 20000 Daltons)

Procedure

Prepare the equilibration buffer solution and filter through a vacuum filter of 0.2 μm pore size. Switch 'ON' AKTA-Prime system and place the buffer solutions A and B. Perform a system wash. Set up Sephadex-G100 or Superdex-200 column and attach to the AKTA-Prime system. Fix the injection valve #1 to column inlet and column outlet to UV monitor flow cell. Attach the sample-loop with 0.5 ml to injection valve at positions "2" and "6". Flush the loop with distilled water using syringe initially and set the injection valve at the "Load" position. Wash the column using 50 ml of distilled water at 0.4 ml/min (pressure at ~0.4 MPa) followed by 50 ml of equilibration buffer.

Void Volume (V_o) determination

Set the injection valve to the "Load" position and apply 250 μL of blue dextran solution into sample-loop using a syringe. Turn the chart recorder 'ON', set the fraction size, switch the injection valve to the "Inject" position, and elute the sample onto the column at a 0.4 ml/min flow rate. A blue band moves down the column first during the elution. Monitor the absorbance at 280-nm during the elution. Set the full-scale absorbance as 0.5. AKTA-Prime settings for manual operation:

Method	Condition
Conc. B	= 0
Gradient	= No (0 ml, 0% B)
Flow rate	= 0.4 ml/min
Fraction size	= 5 ml
Pressure limit	= 1.0 MPa
Buffer valve	= Pos 1
Inject valve	= Inject

Determine the void volume (V_o) from chromatogram by measuring the volume of effluent collected from the point of sample application to the center of the effluent peak. The leading peak indicates the void volume, whereas the trailing edge of the peak shows the heterogeneity of the sample and column packing. It is better to avoid mixing of blue dextran with any of the protein standard because some of the proteins can bind to blue dextran. Prepare the protein standards and blue dextran fresh to minimize the formation of aggregated protein.

Elution Volume (V_e) determination

Dissolve the individual protein standards in equilibration buffer as indicated below:

Protein Standard	Concentration (mg/ml)	Mol. Weight
β-Amylase	8	200,000
Alcohol dehydrogenase	10	150,000
Albumin	10	66,000
Carbonic anhydrase	6	29,000
Myoglobin	2	17,600
Cytochrome C	3	12,400
Aprotinin	2	6,512
Vitamin B12	1	1,355

Mix the standard proteins viz., cytochrome C, β-amylase, carbonic anhydrase, and alcohol dehydrogenase. Set the injection valve at the "Load" position. Apply 250 μl of protein standard into the sample loop using a syringe. Turn the chart recorder "ON", set the fraction size, switch to the "Inject" position, and elute the sample onto the column at a 0.4 ml/min flow rate. Monitor the absorbance at 280 nm during the elution. Set the full-scale absorbance as 0.5.

Determine the elution volume (V_e) for the protein standard by measuring the volume of effluent collected from the point of sample application to the center of the effluent peak. Generate a standard curve by plotting the molecular weight versus V_e/V_0 for each respective protein standard in a semi-log scale.

4.4.4. Separation by Electrophoresis

Electrophoresis relies on the differences in the migration of charged molecules in a solution in an appied electrical field across it. It separates the proteins based on their size, shape, or charge.

4.4.4.1. Non-denaturing Electrophoresis

In non-denaturing electrophoresis, the native proteins are separated based on a combination of their charge, size, and shape. Native proteins dissolved in a buffered solution, are run through a porous gel (polyacrylamide, starch, or agarose), by applying a voltage across it. The movement of proteins through the gel depends on their charge. The rate of movement depends on the magnitude of the charge and the friction.

$$\text{Electrop-oretic mobility} = \frac{\text{Applied voltage} \times \text{molecular charge}}{\text{molecular friction}}$$

Proteins may be positively or negatively charged in solution depending on their isoelectric points (pI) and the pH of the solution. Protein gets negatively charged, if the pH is above the pI and positively charged, if the pH is below the pI. The magnitude of the charge and applied voltage will determine how far proteins migrate at a time. High voltage or greater charge on the protein makes it move further. The friction of a molecule is a measure of its resistance to movement through the gel. The relationship between the size of the molecule and the pores in the gel determines the friction. Smaller size molecules or larger size pores in the gel lower the resistance and move the molecules faster through the gel. Gels with different porosities can be purchased from chemical suppliers or made up in the laboratory. Smaller pore sizes obtained by using a higher concentration of cross-linking reagents to form the gel. Gels may be contained between two parallel plates, or in cylindrical tubes.

4.4.4.2. Denaturing Electrophoresis

In denaturing electrophoresis, proteins are separated primarily on their molecular weight. Proteins are denatured before analysis by mixing them with mercaptoethanol, which breaks down disulphide bonds, and sodium dodecyl sulphate (SDS), which is an anionic surfactant that hydrophobically binds to protein molecules and causes them to unfold because of the repulsion between negatively charged surfactant head-groups. Each protein molecule binds the same amount of SDS per unit length. Hence, the charge per unit length and the molecular conformation is similar for all proteins. As proteins travel through a gel network, they are primarily separated based on their molecular weight, because their movement depends on the size of the protein molecule relative to the size of the pores in the gel. Smaller proteins move more rapidly through the matrix than larger molecules. This type of electrophoresis is sodium dodecyl sulfate-polyacrylamide gel electrophoresis or SDS-PAGE.

A tracking dye added to the protein solution, *e.g.,* bromophenol blue, determines the movement of proteins. This dye is a small charged molecule that migrates ahead of the proteins. After the electrophoresis run, the proteins are made visible by treating the gel with a protein-dye such as Coomassie Brilliant Blue or silver stain. The relative mobility of each protein band calculated as:

$$\text{Relative mobility (Rm)} = \frac{\text{Dis}\tan\text{ce moved by protein}}{\text{Dis}\tan\text{ce moved by dye}}$$

Electrophoresis used to determine the protein composition of food products. The protein is extracted from the food into the solution, which is then separated using electrophoresis. SDS-PAGE is used to determine the molecular weight of a protein by measuring *Rm*, and then comparing it with a calibration curve produced using proteins of known molecular weight: a plot of log (molecular weight) against relative mobility is usually linear. Denaturing electrophoresis is more useful for determining molecular weights than non-denaturing electrophoresis because the friction does not depend on the shape or original charge of the protein molecules.

4.4.4.2.1. SDS-PAGE

Sodium dodecyl sulfate-polyacrylamide gel electrophoresis (PAGE) is one of the most commonly used methods for the molecular weight determination of protein (Laemmli, 1970).

SDS is a charged molecule that makes a charged complex with proteins by binding at the rate of one SDS molecule for every two amino acids. The charge is proportional to the molecular weight of the protein. Dithiothreitol (DTT) or mercaptoethanol reduces disulfide bonds. Denatured proteins when applied to the gel, the electric current causes the negatively changed proteins to migrate across the gel towards the anode. Proteins migrate based on their size and charge. Smaller proteins travel farther down the gel, while larger ones travel a shorter distance. A standard curve constructed using protein markers of known molecular weight. From that the molecular weight of the unknown protein is determined.

Reagents

1. Stock acrylamide: Dissolve 30.0 g of acrylamide and 0.8 g of bisacrylamide in 100 ml of distilled water

2. Separating gel buffer (pH 8.8): Dissolve 22.7 g of Tris – HCl (1.875M) in 100 ml of distilled water

3. Stacking gel buffer (pH 6.8): Dissolve 7.26 g of Tris – HCl (0.6 M) in 100 ml of distilled water

4. Ammonium persulfate, 5% - Prepare fresh before use

5. TEMED – Use fresh from the refrigerator

6. Electrode buffer (pH 8.2 – 8.4): Dissolve 12.0 g of Tris (0.05M), 28.8 g of glycine (0.122M) and 2 g of SDS in 2000 ml of distilled water. Use for 2-3 times.

7. Sample buffer (5X): Dissolve 5 ml of Tris –HCl buffer (pH 6.8), 0.5 g of SDS, 5.0 g of sucrose, 0.25 ml of mercaptoethanol and 1.0 ml of 0.5% bromophenol blue in 10 ml of distilled water and keep frozen.

8. SDS solution, 10%: Store at room temperature

9. Standard marker protein

10. Protein stain solution: Dissolve 0.2 g of coomassic brilliant blue R250 in 40 ml of methanol first and then add 10 ml of acetic acid and 50 ml of distilled water. Prepare the solution fresh every time.

11. De staining solution: Prepare the solution as above but without the dye

Sample preparation

Weigh about 1 g of fish tissue and homogenize with 10 ml of sample buffer, and heat in a boiling water bath for 5-10 min to ensure complete reaction between proteins and SDS. Centrifuge the mixture at 5000xg for 20 min, collect the supernatant, and adjust the protein concentration using sample buffer and water, to a concentration of 50-200mg in a volume of 25-50ml.

Gel polymerization

Clean the glass plates and dry them. Assemble the plates by placing the spacers. Hold the construction together with bulldog clips and clamps in an upright position. Apply white petroleum jelly or 2% agar (melted in a boiling water bath) around the edges of the spacers, to seal the chamber between the glass plates. Prepare required volume of separating gel mixture depending on the size of the glass plates, i.e., 30 ml for a chamber of about 10 x 9 x 0.1 cm size or 10 ml for a chamber 7 x 7 x 0.1 cm size, by mixing as follows:

Reagents	15% gel	12.5 % gel	10% gel
Stock acrylamide solution	20.1 ml	16.7 ml	13.3 ml
Tris – HCl (pH 8.8)	8.0 ml	8.0 ml	8.0 ml
Water	11.4 ml	14.7 ml	18.2 ml
Degas in a water pump for 3-4 min and then add			
Ammonium persulfate solution	0.2 ml	0.2 ml	0.2 ml
10% SDS	0.4 ml	0.4 ml	0.4 ml
TEMED	20μl	20μl	20μl

Mix the gel solution gently and pour carefully in the chamber between the glass plates. Layer the top of the gel with distilled water and leave to set for 30-60 min. Prepare the stacking gel (1%) by mixing the following solutions:

Reagents	Volume
Stock acrylamide solution	1.35 ml
Tris – HCl (pH 6.8)	1.0 ml
Water	7.5 ml
Degas in a water pump for 3-4 min and then add	
Ammonium persulfate solution	50 µl
10% SDS	0.1 ml
TEMED	10 µl

Remove the distilled water from the top of the gel and wash with a little stacking gel solution. Pour the stacking gel mixture, place the comb, and allow the gel to set for 30-60 min. Once the stacking gel has polymerized, remove the comb without distorting the shapes of the well. Install the gel carefully after removing the clips, agar, etc. in the electrophoresis apparatus. Fill the electrode buffer in the chamber. Carefully remove any trapped air bubbles at the bottom of the gel. Connect the cathode to the top (i.e. stacking gel end) and turn the DC power 'ON' to briefly check the electric circuit. Run the electrophoresis in cool condition, to avoid the heat generation that dissipates the run and affect the resolution.

Electrophoretic separation

Load the gel with the sample solution (10-25 µl) using a micropipette carefully into a sample well through the electrode buffer. Mark the position of wells on the glass plate with a marker pen. The presence of bromophenol blue in the sample buffer facilitates the easy loading of the samples. Similarly, load few wells with standard marker proteins prepared using the sample buffer. Turn the current 'ON' to 10-15 mA for an initial period of 5 - 15 min, until the samples travel through the stacking gel and get concentrated as a single streak above the separating gel. Then, continue the run at 30 mA until the bromophenol blue reaches the bottom of the gels. This process takes about 30 min to 3 h. Once the run is complete, carefully remove the gel from the plates and immerse in the staining solution for 3 h or overnight with uniform shaking. The proteins absorb the stain, coomassie brilliant blue. Transfer the gel then to a suitable container containing 200-300 ml destaining solution and shake gently. Remove the dye that is not bound to proteins. Change the destaining solution frequently, particularly during initial periods, until the background of the gel is colorless. Proteins fractionated into the band are colored blue, while proteins of minute quantities stain faintly. Stop the destaining process at an appropriate stage to visualize as many bands as possible. Photograph the gel or store in polythene bags or dry in vacuo for permanent records.

4.4.4.2.2. Isoelectric Focusing

This technique is a modification of electrophoresis, in which proteins separate by the charge on a gel matrix, which has a pH gradient across it. Proteins migrate to the location where the pH equals their isoelectric point and then stop moving when no charge. This method has one of the highest resolutions of all techniques used to separate proteins. Gels are available that cover a narrow pH range (2-3 units) or a broad pH range (3-10 units). One should, therefore, select a gel, which is most suitable for the proteins to be separated.

4.4.4.2.3. Two-Dimensional Gel Electrophoresis

Isoelectric focusing and SDS-PAGE together improve the resolution of complex protein mixtures. Proteins are separated in one direction based on charge using isoelectric focusing, and then in a perpendicular direction based on size using SDS-PAGE.

4.5. CHARACTERIZATION OF PROTEINS AND PEPTIDES

Peptides are formed by enzymatic hydrolysis of proteins or during processing and storage of fish. Many peptides have bioactive properties like immunomodulatory, antihypertensive, antioxidative, and antimicrobial. The term "degree of hydrolysis" describes the extent to which peptide bonds are broken by the enzymatic hydrolysis reaction. It helps to characterize proteins or peptides.

4.5.1. Degree of hydrolysis

Degree of hydrolysis (DH) is a measurement that shows the number of specific peptide bonds broken during hydrolysis as a percent of the total number of peptide bonds present in the intact protein. Several methods are available for the measurement of the degree of hydrolysis. One method determines free amino groups after the reaction with trinitrobenzene-sulfonic acid (TNBS).

In this method, TNBS reacts with un-protonated primary amino acids and form a yellow color product having an absorbance maximum at 340 nm. The reaction takes place under a slightly alkaline condition, later stopped by lowering the pH.

Reagents

1. Sodium phosphate buffer (pH 8.0), 50mM for hydrolysis

    Reagent A – Dissolve 0.069 g of sodium dihydrogen phosphate in 10 ml of distilled water. Reagent B – Dissolve 0.71g of disodium hydrogen phosphate in 100 ml of distilled water. Mix 5.3 ml of reagent A, and 94.7 ml reagent B, adjust the pH to 8.0 and make up the volume to 200 ml with distilled water

2. Trypsin or another proteolytic enzyme

3. SDS,1%

4. Sodium phosphate buffer (pH 8.2), 0.2 M for TNBS reaction

    Reagent A – Dissolve 0.27g of sodium dihydrogen phosphate in 10 ml of distilled water. Reagent B – Dissolve 3.55g of disodium hydrogen phosphate in 100 ml of distilled water. Mix 5.0 ml of reagent A, and 95 ml reagent B, adjust the pH to 8.2 and make up the volume to 200 ml with distilled water

5. Trinitrobenzene-sulfonic acid (TNBS), 0.1%. This a light sensitive compound, so prepare at the time of use.

6. HCl, 0.1N

7. Leucine standard (10 mM): Dissolve 65.5 mg of L-leucine in 100 ml of 1% SDS.

Enzyme hydrolysis

Weigh about 10 g of the tissue sample, and homogenize with 20 ml of 50mM sodium phosphate buffer (pH 8.0) to prepare the homogenate. Take 0.25 ml of homogenate in a test tube containing 2 ml of 1% SDS and incubate at 75°C for 15 min. To the remaining homogenate, add 100 mg of the enzyme (trypsin) and place the mixture in a magnetic stirrer set at 37°C for 2-4 h. At regular time intervals (i.e. once in every 30 min), transfer 0.25 ml of sample into a series of test tubes containing 2 ml of 1% SDS and incubate at 75°C for 15 min to inactivate the reaction.

TNBS reaction

For this reaction, transfer 0.25 ml of sample inactivated by heat before and after the addition of the enzyme into the test tubes containing 2 ml of 0.2M sodium phosphate buffer (pH 8.2). Transfer 0, 0.1, 0.2, 0.3, 0.4, 0.5, and 0.6 ml of leucine standards into a series of test tubes and make up the volume to 1.0 ml with 1% SDS, such that it is equivalent to 0.5, 1.0, 1.5, 2.0, 2.5, and 3.0 amino meq/g of L-leucine, respectively. From that, transfer 0.25 ml of each standard solution into a series of test tubes containing 2 ml of 0.2M sodium phosphate buffer (pH 8.2). Set a blank without the addition of standard and sample, but with 2 ml of 0.2M sodium phosphate buffer (pH 8.2). To all the test tubes, add 2 ml of TNBS reagent, vortex well, cover with an aluminum foil and incubate in dark at 50°C for 60 min. Stop the reaction with the addition of 4 ml of 0.1N HCl. Allow the test tubes to cool at room temperature and read the absorbance at 340 nm in a spectrophotometer. Construct a calibration graph by plotting the concentration of L-leucine in the X-axis and the absorbance value on the Y-axis. Compute the L-leucine as amino meq/g in the sample from the graph.

Calculation

$$\text{Degree of Hydrolysis, \%} = \frac{C_t - C_o}{C_o} \times 100$$

where, C_o – L-leucine equivalent at 0 time; C_t – L-leucine equivalent at regular time intervals

4.5.2. Peptide fractionation

Peptides give valuable information on the quality of the food, especially the enzymes that are active during storage. Selective precipitation, membrane ultrafiltration, and gel filtration are some techniques used for the determination of peptide fractions.

4.5.2.1. Selective precipitation

Solvents of different polarities induce protein and peptide precipitation. For the determination of the number of peptides below a chain length, use selective precipitation using ethanol, methanol, or trichloroacetic acid. Solvent precipitation is not suitable because their residues are undesirable for food and nutraceutical industries. However, ethanol precipitation is acceptable due to its volatile nature and ease of removal by evaporation. The amounts of peptides soluble in different concentrations of ethanol are dependent on the chain length as well as on the hydrophobicity of the peptides. Protein precipitation helps to study peptides of lower concentrations by chromatographic methods such as LC-MS or electrophoretic methods. Mass spectroscopy determines the molecular mass of the peptides and tandem mass spectroscopy to provide detailed information of the structure of the peptides.

4.5.2.2. Membrane ultrafiltration

Membrane-based processes like ultrafiltration (UF) and nanofiltration (NF) can fractionate proteins and peptides. Dissolve 0.1g of lyophilized protein hydrolysate in 10 ml of 0.1 M sodium phosphate buffer (pH 7.0). Use the liquid protein hydrolysate directly. Centrifuge the mixture at 9,000xg for 20 min at 4°C and filter through 0.45µm filter. Transfer the filtrate to a 30kDa molecular weight cut off (MWCO) ultrafilter and centrifuge at 9000xg for 30 min. Sequentially transfer the filtrate to ultra-filters with MWCO of 10, 5, 3, and 1 kDa and centrifuge at 9000xg for 15 min. Lyophilize each filtrate fraction with different MWCO and store at -40°C in aluminum foil bags.

4.5.2.3. Gel filtration

Sephadex G-25 column can separate peptide fractions. First, concentrate the sample using rotary flash evaporator to a thick solution, and solubilize with 5% formic acid, before injection into the column chromatography. Run at 0.2 ml/min flow rate using 5% formic acid and monitor the effluent at 280 nm. Collect 4ml fractions using fraction collector until no peaks observed. Pool the fractions according to elution profile, lyophilize and store at -40°C in aluminum foil bags. Determine the molecular weight of each fraction by SDS-PAGE.

4.5.3. Analysis of protein modifications

Muscle proteins are vulnerable to oxidative attacks during the processing and storage of muscle foods. Oxidation can occur at both the protein backbone and on the amino acid side chains resulting in physical changes in protein structure ranging from the fragmentation of the backbone to the oxidation of the side chains. Oxidation of protein side chains gives rise to unfolding and conformational changes in protein, leading to dimerization or aggregation. Oxidative modification also leads to alterations in the functional, nutritional, and sensory properties of the muscle proteins, including gelation, emulsification, viscosity, solubility, and water holding capacity. Several methods can determine protein oxidation. The most commonly used methods relate to carbonyl group formation, SH groups reduction, and di-tyrosine formation.

4.5.3.1. SH group determination by DTNB method

Peptides containing free sulfhydryl groups react with Ellman's reagent, 5,5'-dithiobis (2-nitrobenzoic acid) [DTNB] and produce a yellow-colored p-nitro thiophenol having an absorption maximum at 412nm.

Reagents

1. EDTA, 0.02 M
2. Tris- EDTA buffer: Dissolve 4.84g of tris, in 100 ml of distilled water. Dissolve 1.49 g of EDTA in 20 ml of distilled water. Combine both the solutions and make up to 200ml with distilled water
3. DTNB (dithio bis-2-nitrobenzoic acid) reagent, 0.01M
4. L- Cysteine HCl standard

   Stock (1mM/ml): Dissolve 17.56 mg of L- cysteine HCl, in 100 ml of distilled water

   Working I (10µM/ml): Dilute 1 ml of the Stock to 100 ml with distilled water

   Working II (0.2mM/ml): Dilute 1 ml of working I to 50 ml with distilled water

Procedure

Prepare a protein solution in duplicate at a concentration of 1-5mg/ml with distilled water in test tubes. Take L-cysteine HCl standards in a series of test tubes to obtain concentration from 0.2 to 10mM/ml. To all the test tubes, add 1.5 ml of 0.2M Tris buffer, pH 8.2 and 0.1ml of 0.01M DTNB reagent and make up the volume to 10 ml with 7.9 ml of absolute methanol. Prepare a reagent blank (without sample), and a sample blank (without DTNB), in duplicate in separate test tubes. Cover the test tubes, allow to stand with occasional shaking for 30min, and filter. Read the absorbance (O.D) of the clear filtrate at 412 nm in a spectrophotometer. Construct a calibration graph with the absorbance versus L-cysteine concentration. Determine the L-cystine present in the sample solution from the calibration graph.

Calculation

$$\text{Sulfhydryl groups} = \frac{C \times 1000}{W}$$

where, C – conc. of cysteine, µM; 1000 – mg to g conversion; W – weight of protein in 1 ml, mg

4.5.3.2. Tyrosinase activity

Tyrosinase activity is the most characteristic test for measuring the primary cleavage in the protein.

Primary cleavage of protein yields tyrosine either in free form or peptide form, which reacts with molybdic phosphoric reagent to give blue colored tyrosine reaction complexes. The complex includes other phenols, tryptophan, cysteine, sulfhydryl compounds (H₂S), and other reducing agents having an absorption maximum at 660 nm.

Reagents

1. NaCl, 0.15 N

2. Trichloroacetic acid, TCA, 10% and 2.5%

3. NaOH, 0.1 N

4. **Tyrosine standard**

   Stock (12.5mg/ml): Dissolve 1.25g of L-tyrosine, in 100 ml of 2.5% TCA

   Working (125µg/ml): Dilute 1 ml of the stock to 100 ml with 2.5% TCA

5. **Reagent A:** Dissolve 2 g of sodium carbonate in 100 ml of 0.1 N NaOH

6. **Reagent B:** Dissolve 50 mg of copper sulfate in 10 ml of 1% sodium potassium tartrate

7. **Reagent C:** Mix reagent A, 50 ml and reagent B, 1 ml just before use

8. **Folin Ciocalteau reagent:** Use readymade reagent

Procedure

Weigh 10 g of the tissue sample, homogenize with 50 ml of chill 0.15 M NaCl, and filter it through moist gauze. Incubate the filtrate approximately 6 ml for 1 h at 40°C in a water bath. Stop the enzyme reaction with the addition of 12 ml of 10% TCA. Incubate the mixture at 4°C for 15 min and centrifuge at 6000xg for 15 min at 4°C. Transfer 0.1 and 0.5 ml of the supernatant into two test tubes and make up the volume to 1 ml with 2.5% TCA. Transfer 0.2, 0.4, 0.6, 0.8, and 1.0 ml of tyrosine standards in a series of test tubes and make up the volume to 1 ml with 2.5% TCA to give concentrations of 25, 50, 75, 100, and 125 µg, respectively. Prepare a blank with 1 ml of 2.5% TCA in a test tube. To all the test tubes, add 2.5 ml of Reagent C followed by 0.5 ml of Folin Ciocalteau reagent and leave for 30 min in dark. Read the absorbance (O.D) at 660 nm in a spectrophotometer. Prepare a standard graph by plotting tyrosine concentration in the X-axis and absorbance value in the Y-axis. Calculate the tyrosine concentration present in the sample from the graph.

Calculation

$$\text{Tyrosinase activity} = \frac{C \times D.F}{W \times 1000}$$

where, C – concentration of tyrosine, μg; D.F – dilution factor; W – weight of the fish, g; 1000 - μg to mg conversion

4.5.3.3. ATPase activity

Myofibrillar ATPase activity distinguishes fast and slow contracting muscle fibers during the cross-bridge cycle, as myosin molecule binds itself and hydrolyzes ATP during force generation. Myosin ATPase activity correlates positively with muscle contraction velocity. Fast-contracting muscle fibers hydrolyze ATP faster than slow-contracting fibers. The first step is pre-incubation of muscle tissue with either an acid (pH ~4) or basic (pH~10) solution. Acid pre-incubation inhibits the myosin ATPase activity in fast (Type 2A, 2X, and 2B) fiber types, but slow in Type 1. Conversely, basic pre-incubation inhibits myosin ATPase in slow fiber types only.

In this assay, ATP is the reaction energy source and substrate, Pi is the reaction product, and myosin is the enzyme. The reaction product Pi released by the myosin molecules by consumption of ATP is invisible. The Pi chemically reacts with calcium to form the precipitate calcium phosphate (CaPO$_4$, limestone) and gets converted into a less soluble brownish-black product cobalt sulfide (CoS$_2$) that deposit on the muscle tissue.

Reagents

1. NaCl, (pH 7.2), 0.6M
2. Tris malate buffer (pH 7.0), 0.5M: Dissoleve 11.86 g of tris malate in 100 ml of distilled water.
3. Trichloroacetic acid (TCA), 15%
4. Calcium chloride, 0.1M
5. Magnesium chloride, 0.02M
6. EDTA, 0.005 M
7. **Sodium ATP solution, 20mM:** Dissolve 5.5 mg of sodium ATP in 100 ml of distilled water.

8. **Amino napthal sulphonic acid reagent (ANSA):** Prepare 15% of sodium bisulphate solution first. Take 195ml of this solution in a beaker and add 0.5g of 1,2,3 amino naphthol sulphonic acid, and 5ml of 20% sodium sulphite solution, mix well, filter and store in a brown bottle.

9. **Molybdate solution:** Dissolve 2 g of ammonium molybdate tetrahydrate $(NH_4)_6Mo_7O_{24}.4H_2O$ in 50 ml of hot distilled water and cool.

10. **Metavanadate solution:** Dissolve 0.2 g of ammonium metavanadate (NH_4VO_3) in 25 ml of hot distilled water, cool and then add 25 ml of 70% $HClO_4$.

11. **Molybdovanadate reagent:** Mix 20 ml of molybdate solution and 25 ml of metavanadate solution and dilute to 100 ml with distilled water.

12. **Potassium orthophosphate monobasic (KH_2PO_4) standard**

    Stock (2 mg P/ml): Dissolve 0.8788 g of KH_2PO_4 in 100 ml in distilled water

    Working (0.1 mg P/ml): Dilute 0.5 ml of the stock to 100 ml with distilled water

Extraction of actomyosin

Weigh 5 g of the tissue sample, homogenize well for 2 min with 45ml of chilled NaCl buffer solution (0.6M NaCl, pH 7.2), and centrifuge at 9,000xg for 20 min at 4°C. Dilute the supernatant in 2 volumes of chilled distilled water and place it in the refrigerator overnight, to settle the actomyosin. Separate the precipitated actomyosin by centrifuging at 9000xg for 20 min at 4°C. Re-constitute the precipitate in 30 ml of chilled 0.6M NaCl (pH 7.2) (Jiang *et al.*, 1987).

Ca – ATPase activity

Take 1 ml of actomyosin solution containing 1-5mg /ml protein in a test tube and add 0.5ml of 0.5M tris maleate buffer (pH 7.0), 0.5ml of 0.1M calcium chloride, 5 ml of distilled water, and 0.5ml of 20mM disodium ATP solution. Estimate the rate of release of inorganic phosphate at 25° C within 3min of reaction. Then, add 5 ml of 15% TCA to stop the reaction and estimate the inorganic phosphate released. Determine the calcium ATPase specific activity as milli mole of C- PO_4 released for mg of actomyosin within 1 min of reaction at 25°C. Express the calcium ATPase total activity as milli mole of PO_4 released per min for actomyosin from 5g of fish.

Mg- ATPase activity

Take 1 ml of actomyosin solution containing 1 –5mg /ml protein in a test tube and add 1ml of 0.02M magnesium chloride, 1ml of 0.005 M EDTA, 1ml of 0.2M tris malate buffer (pH 7.0), 5ml of distilled water, and 1 ml of 20mM disodium ATPase solution. Estimate the rate of release of inorganic phosphate at 25°C within 3min of reaction after the addition of ATP. Calculate the Mg- ATPase specific activity and Mg-ATPase total activity.

Inorganic phosphorus estimation

Transfer 1ml of the reaction mixture in a test tube, add 2 ml molybdovanadate reagent and make up the volume to 10ml with distilled water. Transfer 0.5, 0.8, 1.0, 1.5, and 2.0 ml of potassium orthophosphate monobasic (KH_2PO_4) standards in a series of test tubes and make up the volume to 10ml with distilled water. Add 2 ml of molybdovanadate reagent. Leave for 10 min and measure the absorbance (O.D) at 400 nm in a spectrophotometer. Construct a calibration graph with the absorbance versus orthophosphate (mg P) content and determine the phosphate content (mg P) in the sample solution from the graph.

4.5.4. Analysis of functional properties of proteins

Protein oxidation results in loss of functional properties such as solubility, water holding capacity, gelling and emulsification, and formation of aggregates. Functional properties fall into three groups according to the mechanism of action.

1. Properties related with hydration – absorption of water or oil, solubility, thickening, and wettability
2. Properties related with the protein structure and rheological characteristics – viscosity, elasticity, adhesiveness, aggregation, and gelation
3. Properties related with the protein surface – emulsifying and foaming activities, formation of protein-lipid films, and whipping ability

4.5.4.1. Solubility

Solubility is the ability of the protein to dissolve in different ionic media. Major interactions that influence the solubility of proteins are hydrophobic and ionic interactions and other conditions such as pH, ionic strength, temperature, and presence of organic solvents. Solubility of proteins measures the extent of denaturation during extraction, isolation, and purification processes. Solubility characteristics expressed as protein solubility index (PSI) or protein dispersibility index (PDI) as the percentage (%) of soluble protein present in a protein sample. PSI of commercial protein isolates varies from 25% to 80%. The procedure for the determination of solubility of protein is given (Chawla *et al.* 1996).

Reagents

1. Solution 1 - 0.6 M KCl
2. Solution 2 - 20 mM Tris-HCl
3. Solution 3 - 20 mM Tris-HCl containing 1% SDS
4. Solution 4 - 20 mM Tris-HCl containing 1% SDS and 8M urea
5. Solution 5 - 20 mM Tris-HCl containing 1% SDS, 8M urea and 2% β-mercaptoethanol

Procedure

Weigh 1 g of any protein and homogenize in 10 ml of the solution (S1-S5) individually. Leave the solution at room temperature for 4 h and centrifuge at 7500xg for 30 min. Collect the supernatant and estimate the protein content by Biuret or Kjeldahl method. Report the solubilized protein to that of the total protein as solubility, in percentage.

4.5.4.2. Water holding capacity and Fat binding capacity

Water binding or hydration capacity of protein is grams of water-bound per gram of protein equilibrated with water vapor at 90-95% relative humidity. The water holding capacity of a protein is important than the water-binding capacity in food applications. Water holding capacity (WHC) refers to the ability of the protein to imbibe water and retain it against gravitational force within a fish muscle. This water is the sum of bound water, hydrodynamic water, and physically entrapped water. Physically entrapped water contributes more to the water holding capacity. The ability of proteins to entrap water is associated with juiciness and tenderness of fish products. Fat binding capacity (FBC) or Fat absorption capacity (FAC) or Fat holding capacity (FHC) is a measure of the amount of oil absorbed per g of protein. The procedure for the determination of WHC and FBC of protein is given (Cho *et al.*, 2004).

Reagents

1. Sunflower oil
2. Distilled water

Procedure

Weigh 10 mg of any protein in a centrifuge tube containing 0.5 ml distilled water for testing WHC or 0.1 ml of sunflower oil for testing FBC. Leave it at room temperature 1h for binding and then vortex it for 5 sec at a time interval of 15 min for occasional mixing. Centrifuge at 4500xg for 20 min and remove the supernatant by draining the tube for 30 min on a filter paper by tilting to a 45° angle.

Calculation

$$\text{WHC or FBC (g/100 g)} = \frac{\text{Initial wt. of the tube} - \text{Wt. of the tube after draining}}{\text{Wt. of the dried protein, g}}$$

4.5.4.3. Viscosity

The viscosity of the solution relates to its resistance to flow under an applied force or shear stress. The flow behavior of the solution gets influenced by the solute type. There are many ways to express the viscosity of dilute protein solutions. Relative viscosity refers to the ratio of the viscosity of the protein solution to that of the solvent. Oswalt-Fenske type capillary viscometer measures the relative viscosity.

$$\text{Relative viscosity } (\eta\text{rel}) = \frac{\eta}{\eta_0} = \frac{rt}{r_0 t_0}$$

where r and ro are the densities of protein solution and solvent, respectively, and t and to are the time taken for the flow of a given volume of protein solution and solvent, respectively, through the capillary. Intrinsic viscosity (η) is obtained by extrapolating a plot of reduced viscosity versus proteins concentration to zero protein concentration. There are two methods to determine viscosity: Ostwald's viscometer and Brookefield viscometer.

4.5.4.3.1. Oswald's Viscometer

Prepare a 0.1% protein solution in a suitable solvent (myofibrillar protein in a salt solution, or collagen in 0.1 M acetic acid). Fill the Ostwald's viscometer with the protein solution up to the mark. Take the respective solvent (distilled water or acetic acid) as a control in another Ostwald's viscometer. Place the viscometers in a water bath set at 30°C. Measure the flow times of the sample and control using a stopwatch and calculate the average flow rates as indices for the relative viscosities (Sivakumar *et al.*, 2000).

Calculation

$$\text{Relative viscosity } (\eta_{rel}) = \frac{\text{Flow time of sample}}{\text{Flow time of control}}$$

$$\text{Specific viscosity } (\eta_{sp}) = \eta_{rel} - 1$$

4.5.4.3.2. Brookfield's viscometer

Prepare appropriate concentration of protein solution in distilled water (1% in case of myosin or 6.67%, if for gelatin). Transfer the solution into a 250 ml beaker and place it in Brookfield Digital Viscometer equipped with a No. 1 Spindle that rotates at 60 rpm at ambient temperature (30±2°C). Read the viscosity (in cP) in the viscometer (Cho *et al.*, 2006).

4.5.4.4. Bloom strength

Protein gelation refers to the transformation of a protein from the 'sol' state to a 'gel-like' state. Heat, enzymes, or divalent cations under appropriate conditions facilitate this transformation. Interactions involved in gel network formation are primarily hydrogen bonds, hydrophobic and electrostatic interactions. Protein forms two types of gel: coagulum (opaque) gels and translucent gels. Type of gel formed by a protein is due to its molecular properties and solution conditions. Proteins containing large amounts of non-polar amino acid residues undergo hydrophobic aggregation upon denaturation. These insoluble aggregates randomly associate and set into an irreversible coagulum-type gel. Proteins that contain small amounts of non-polar amino acid residues form soluble complexes upon denaturation. These translucent gels formed primarily by hydrogen bonding interaction hold more water than coagulum type gels. The protocol to determine the bloom strength of the protein gel is given (Cheow *et al.* 2007).

Procedure

Prepare appropriate concentration of protein solution (1% in case of myosin or 6.67%, if for gelatin) with distilled water in a beaker of a dimension of 4.8 cm x 1.8 cm. Heat the solution at 65°C for 25 min (in case of gelatin), cool to room temperature, and allow to mature at 4°C for 16-18 h. Determine the bloom strength in a Universal Testing Machine or Texture Analyzer, equipped with a 12.7 mm dia flat-face cylindrical Teflon-coated plunger at a cross-head speed at 0.5 mm/s. Record the maximum force (g) as the plunger penetrates 4 mm into the gel.

4.5.4.5. Emulsification

There are many methods to evaluate the emulsifying properties of food proteins: size distribution of oil droplets, emulsifying activity, emulsion capacity, and emulsion stability. The protocol to determine the emulsion stability index (ESI) and emulsifying activity index (EAI) of the protein by the turbidimetric method is given (Pearce and Kinsella, 1978).

Reagents

1. Phosphate buffer 10 mM (pH 7.0)

    Reagent A: 1M K_2HPO_4. Dissolve 17.418 g of K_2HPO_4 in 100 ml of distilled water.

    Reagent B: 1M KH_2PO_4 Dissolve 13.609 g of KH_2PO_4 in 100 ml of distilled water.

    Mix 6.15 ml of reagent A and 3.85 ml of reagent B and dilute to 100 ml with distilled water.
2. Sunflower oil
3. Sodium dodecyl sulfate (SDS), 0.1%

Procedure

Prepare 1% protein solution in 10 mM phosphate buffer (pH 7.0). Take 4.5 ml of the solution in a beaker and add 1.5 ml of vegetable oil. Homogenize the mixture for 1 min in a magnetic stirrer. Take 50 µl immediately after homogenization (0 min) and after 15 min in two separate volumetric flasks and make up the total volume to 5 ml with 0.1% of SDS solution. Read the absorbance at 500 nm in a spectrophotometer.

Calculation

$$\textit{Emulsion stability index (ESI) in min} = \frac{A_0 \times t}{A_0 - A_{15}}$$

where, A_0- absorbance at time 0; A_{15}-absorbance at time 15 min; t- time interval (15 min)

$$\textit{Emulsifying activity index (EAI)} \text{ in } m^2/g = \frac{2 \times T \times A_0 \times D.F}{C_{xx}}$$

where, T - 2.303 constant; A_0 absorbance at time 0; D.F – dilution factor; C – Wt. of protein per unit volume (g/ml); x - Vol. of oil fraction taken (50 μlor 0.05 ml);

4.5.4.6. Foaming properties

Foam has a continuous aqueous phase and a dispersed gaseous (air) phase. Foaming property is the ability of a protein to form a tenacious film at the gas-liquid interface to incorporate and stabilize large quantities of gas bubbles. There are two major ways to evaluate the foaming properties: foamability and foam stability. Foamability or foaming capacity of a protein refers to the amount of interfacial area created by the protein expressed as foaming power. Foam stability refers to the ability of protein to stabilize foam against gravitational and mechanical stresses and is often expressed as the time required for 50% of the liquid to drain from foam or for a 50% reduction in foam volume. The most direct measure of foam stability is the reduction of foam interfacial area as a function of time. The protocol to determine the foaming properties of protein is given (Cheow *et al.* 2004).

Procedure

Prepare a 1% protein solution in 100 ml of distilled water and heat at 60°C for complete solubilization. Note down the initial volume after solubilization (VI_0). Homogenize the solution continuously in a magnetic stirrer for 30 min to produce foam. Pour the foam solution into a 100 ml measuring jar and immediately measure the volume of foam (VF_0) and volume of liquid (VL_0). Then, measure the volume of foam (VF_{30}) after 30 min, to calculate the foam stability.

Calculation

$$\text{Foam ability (FA)} = \frac{VF_0 + VL_0}{VI_0}$$

$$\text{Foam stability (FS)} = \frac{VF_0 - VF_{30}}{VF_0}$$

4.5.5. Analysis of amino acids

There are more than 100 different amino acids isolated from biological materials, but only 25 present in proteins. Amino acids have an acidic carboxyl (-COOH) and a basic nitrogenous amino group ($-NH_2$). They serve as a metabolic source of energy. Amino acids are also essential for carbohydrate and lipid metabolism and synthesis of tissue proteins, and other compounds such as adrenalin, thyroxine, melanin, histamine, porphyrins, pyrimidines, purines, choline, folic acid, nicotinic acid, taurine, etc. There are two groups of amino acids from nutritional point of view: essential amino acids (EAA) and non-essential amino acids (NEAA). Dietary EAA in fish and shrimp are threonine, leucine, methionine, lysine, arginine, valine, isoleucine, tryptophan, histidine, and phenylalanine. Amino acids are not only building blocks of protein but also primary constituents, or nitrogen precursors for many non-protein nitrogens containing compounds.

The amino acid analyses quantify protein and peptides, determine the identity of proteins or peptides based on their amino acid composition, support protein/peptide structure analysis, evaluate fragmentation strategies for peptide mapping and detect atypical amino acids present in a protein or peptide. The amino acid analysis is similar to the determination of free amino acids for these analyses.

4.5.5.1. HPLC method for total amino acids

A highly reactive amine, 6-aminoquinolyl-N hydroxysuccinimidyl carbamate (AQC) reacts with amino acids to form stable urea derivatives for further analysis by reversed-phase HPLC. Primary and secondary amino acids derivatize quickly and remain stable for more than 7 days at room temperature.

Reagents

1. HCl, 6 N and 0.1 N
2. Internal standard: α-aminobutyric acid 2.5 µmol/ml (AABA)
   Stock (50µmol/ml): Dissolve 0.258 g of AABA to 50 ml of 0.1N HCl.
   Working (2.5µmol/ml): Transfer 1 ml of the stock in a 20 ml volumetric flask and make up to volume with 0.1N HCl
3. AccQ•Fluor borate buffer
4. AccQ•Fluor reagent (6-aminoquindyl-N hydroxy succinimidyl carbamate)
5. Pierce H amino acid standards
6. Eluent A – Acetate phosphate buffer
7. Eluent B – 60% Acetonitrile

Acid hydrolysis

Weigh 30-50 mg of the dried sample in a hydrolysis tube depending on the protein concentration. Add 5 ml of 6 N HCl, purge with nitrogen gas for 30 sec, and seal immediately. Place the tube in an electric oven for 24 h at 110°C for hydrolysis and cool. Add 400 μl of internal standard, α-aminobutyric acid, and transfer to a 100 ml volumetric flask and dilute to volume with distilled water. Filter 1 ml and transfer 10 μl of the filtrate to a small derivatization tube (6 × 50 mm).

Derivatization

To the filtrate, add 70 μl of AccQ•Fluor borate buffer using a micropipette and vortex for a few sec. Then, add 20 μl of AccQ•Fluor reagent and vortex for a few secs. Allow the mixture to stand for 1 min at room temperature and heat the sample at 55°C for 10 min in a water bath. Transfer the entire content to the auto-sampler capped with a silicone-lined septum.

Preparation of internal standard and calibration standard

Use the internal standard, AABA stock to prepare the calibration standard. Calibration standard consists of a 1:1 (v/v) mixture of amino acid standard, which contains 2.5 μmol/ml of each amino acid standard, except for 1.25 μmol/ml of cysteine, and a 2.5 μmol/ml of AABA. Prepare a calibration standard with an internal standard by combining 80 μl of 2.5 μmol/ml AABA (working) with 80 μl of amino acid standard and make up the volume to 1000 μl with 840 μL of HPLC grade water. Calibration standard contains 2 nmoles of each standard amino acid component from 1:1 (v/v) mixture of amino acid standard and 2.5 μmol/ml of AABA. Transfer 10 μl of calibration standard in a derivatization tube, add 70 μl of AccQ•Fluor borate buffer using a micropipette with vortexing and add 20 μl of AccQ•Fluor reagent with vortexing. Allow the mixture to stand for 1 min at room temperature and heat the sample at 55°C for 10 min in a water bath for derivatization. Transfer the entire content to the auto-sampler capped with a silicone-lined septum.

HPLC analysis

This protocol explains the analysis by the Waters AccQ•Tag amino acid analyzer fitted with fluorescence detector. Set the excitation wavelength at 285 nm, and the emission wavelength at 354 nm. Set the filter and gain as 1.5 sec and 10, respectively. Eluent A and eluent B are acetate-phosphate buffer and 60%

acetonitrile in water, respectively. Set the column temperature at 37°C. Condition the column first with eluent B at 1 mL/min flow rate for 5 min, followed by equilibration with 100% AccQ•Tag eluent A for 9 min at the same flow rate. Keep the consistent period of the equilibration for all the analyses. Run a blank with Milli-Q water or HPLC grade water before each analysis to determine baseline performance. Follow a gradient programming, for the analysis of amino acids as given below. Set the total run time as 50 min. Determine the concentration of amino acids based on the area percentage of the individual amino acid in the sample with that of the standard.

Time (min)	Flow rate (ml/min)	%A	%B
Initial	1.0	100.0	0.0
0.5	1.0	98.0	2.0
15.0	1.0	90.0	10.0
19.0	1.0	87.0	13.0
32.0	1.0	65.0	35.0
34.0	1.0	65.0	35.0
37.0	1.0	0.0	100.0
38	1.0	100.0	0.0
50.0	1.0	100.0	0.0

4.5.5.2. Available lysine estimation

Heat processing of fish destroys certain amino acids, particularly, lysine affects the nutritive value of protein. The 'available' amino acid is distinct from 'total' amino acid present in any protein. The 'available' amino acid differentiates between amino acids modified or damaged during denaturation with the loss of nutritive value, and those that remain nutritionally available to the metabolic processes. Heat denaturation causes amino acids to decompose and to react chemically with other compounds, if reactive functional groups, such as the ε-amino group of lysine, is present at reactive sites. A chemical method for the estimation of available lysine depends on the reactivity of the ε-amino group with 1-fluoro-2,4-dinitrobenzene, to form mono ε-2,4-dinitrophenyl lysine (Carpenter, 1960).

Free ε-amino groups of lysine in protein first react with fluoro dinitrobenzene (FDNB) at alkaline pH 8.0 and get hydrolyzed with HCl. The DNP-lysine molecules eliminated from the hydrolysates reacts with methoxycarbonyl chloride and extracted by the ether. The resultant hydrolysate has an absorption maximum at 435 nm.

Reagents

1. Sodium bicarbonate, 8%

2. 1-Fluorodinitro benzene (FDNB), 2.5% in ethanol

3. Ethanol, 95%

4. Hydrochloric acid, 8.1N

5. Hydrochloric acid, 1 N

6. Sodium hydroxide solution, 2N

7. Buffer, pH 8.5: Mix 19 parts of 8% $NaHCO_3$ and 1 part of 8% Na_2CO_3; adjust the pH with HCl or NaOH, if needed

8. Methoxy carbonyl chloride

9. Dinitro phenyl lysine (DNP - lysine) standard

   Dissolve 25 mg of DNP – lysine HCl in 500ml of 1N HCl. This solution contains 39.85 mg lysine/ml or 50mg of DNP lysine / ml.

Procedure

Weigh accurately 0.5±0.05g of the dried sample in a round-bottomed flask, add 8 ml of 8% $NaHCO_3$ and shake for 10 min. Then, add 0.3 ml of FDNB solution prepared in 2ml ethanol. Shake continuously but gently 2h. Evaporate the ethanol off by placing it in a boiling water bath. Then, add 24 ml of 8.1 N HCl to the mixture and reflux gently 6 h, cool and filter the content. Wash with water and make up the filtrate to 250 ml with distilled water. Dilute the filtrate if necessary, such that 2 ml of the filtrate contains about 50 µg of lysine. Transfer 2ml each of diluted filtrate to two separate stopper test tubes A and B; and another 2 ml to a conical flask C. Make up the volume of tube A to 10 ml with 1N HCl and read the absorbance at 435nm in a spectrophotometer (A). Titrate the content in flask C with 2N NaOH using phenolphthalein as an indicator. Add the same titer volume of 2N NaOH to the tube B to neutralize it. Then, add 2 ml of buffer (pH 8.5) and 0.5ml of methoxycarbonyl chloride and shake well. After 5–10 min, add 0.75ml of conc. HCl with agitation to it. Extract the content twice with 5 ml portions of ether, discard the ether layer, evaporate off the residual ether by placing in boiling water and make up the volume to 10 ml with distilled water. Read the absorbance at 435nm in a spectrophotometer (B) and subtract reading (A – B) to get the absorbance of ε-DNP lysine. Compare the values with 2 ml of standard DNP lysine processed through the same procedure in Tube A, except ether washing. Calculate the equivalent amount of lysine with a suitable correction factor of 1.09 for the losses occured due to hydrolysis.

Calculation

$$\text{Available lysine, (mg/100 g)} = \frac{250}{W} \times \frac{C}{2} \times 1.09 \times 100$$

where, W- weight of fish, g; C – conc. of lysine, mg; 1.09- correction value; 100- % conversion

4.5.5.3. Collagen estimation

Collagen is the major connective tissue protein present in the fish. Hydroxyproline and proline are the two amino acids in the collagen found in larger amounts. Hydroxylysine is exclusively found in collagen and hence collagen determined by analysis of hydroxyproline. The amount of hydroxyproline residues per 100 residues in the collagen is known for an accurate determination of the collagen, as their quantity varies for different collagen types, their parts, and species.

Tissue sample hydrolyzed in sulfuric acid at 105°C to liberate the amino acids. Hydroproline oxidizes with chloramine-T to pyrrole, and with the addition of 4-dimethyl amino benzaldehyde (DMAB) develops a red-purple color compound having an absorbance maximum at 558 nm.

Reagents

1. Sulfuric acid, 3.5 M
2. **Citrate buffer (pH 6.0)**

    Dissolve 30 g of citric acid monohydrate, 15 g of sodium hydroxide and 90 g of sodium acetate trihydrate in 500 ml of distilled water and then add 290 ml of n-propanol and check the pH. Transfer the content to a 1L volumetric flask and dilute to volume using distilled water. This solution is stable for 2 months at 4°C in a dark bottle.

3. **Oxidant solution**

    Dissolve 1.41 g of chloramine-T in 100 ml of citrate buffer. This solution is stable for 1 wk at 4°C in a dark bottle.

4. **Color reagent**

    Dissolve 10 g of dimethyl amino benzaldehyde (DMAB) in 35 ml of perchloric acid ($HClO_4$) and then add slowly 65 ml of 2- propanol. This solution is prepared fresh every day.

5. **Hydroxyproline standard**

Stock (600 μg/ml) – Dissolve 60 mg of L-hydroxyproline in 100 ml of distilled water. This solution is stable for 2 months at 4°C.

Intermediate (6 μg/ml)– From the stock, dilute 5 ml with 500 ml of distilled water. This solution is prepared fresh every day.

Working standard– From the intermediate, dilute 10, 20, 30, and 40 ml taken in into a series of 100 ml volumetric flasks with distilled water. Each working standard solution contains 0.6, 1.2, 1.8, and 2.4 μg of hydroxyproline/ml, respectively. These solutions are prepared fresh every day.

Procedure

Weigh accurately 5±0.5 g of fish skin or scale or gills in a conical flask and add 30 ml of H_2SO_4, cover with watch glass, and place at 105°C for 16 h for hydrolysis. Transfer the hot hydrolysate into 500 ml volumetric flask and dilute to volume with distilled water. Filter a part of the solution through Whatman No. 1 filter paper and dilute 5 ml of the filtrate into 100 ml of distilled water, such that it contains hydroxyproline ranging from 0.5-2.4μg/ml.

From the filtrate, transfer 2.0 ml into a test tube and add 1.0 ml of oxidant solution, shake well, and allow to stand at room temperature for 20 min. Then, add 1.0 ml of color reagent, mix thoroughly, cover, and place it in a water bath set at 60°C for exactly 15 min. Read the absorbance at 558 nm in a spectrophotometer. Transfer 2.0 ml of distilled water in a test tube to serve as a blank. Transfer 2.0 ml of each working standard into a series of test tubes and repeat the procedure as that of the sample. Draw a standard graph by plotting the concentration of hydroxyproline on the X-axis and O.D on the Y-axis. Determine the concentration of the hydroxyproline in the sample from the graph.

Calculation

$$\text{Hydroxyproline, \%} = \frac{C \times D.F \times 100}{W \times 10^6}$$

where, C – hydroxyproline, μg in 2 ml filtrate; W – weight of test portion, g; D.F - dilution factor; 10^6 - μg to g conversion factor, 100 – percentage conversion

$$\text{Collagen, \%} = \text{Hydroxyproline \%} \times 8$$

Note: Collagenous connective tissue contains 12.5% hydroxyproline, if nitrogen-to-protein factor is 6.25

$$\text{Collagen per unit crude protein, \%} = \frac{\text{Collagen, \%} \times 100}{\text{Crude protein, \%}}$$

CHAPTER 5

ANALYSIS OF VITAMINS

Vitamins are a group of complex organic compounds essential for the normal functioning of metabolic reactions in the body. There are two categories based on their solubility: fat-soluble and water-soluble vitamins. The concentrations of vitamins present in fish and seafood available in the US Department of Agriculture National Nutrient Database (http://www.ars.usda.gov/nutrientdata), along with the percentages of the daily value (DV) of vitamins. Foods containing 20% or more of the DV of nutrients per reference amount are the excellent sources of nutrients. Foods containing 10-19% of the DV are categorized as good sources. All the vitamins necessary for good health are present to some extent in fish. The vitamin content of fish varies within the same species, their parts, and even season. Vitamins A and D are abundant in the liver of lean fish and shellfish and the meat of fatty fish. Fish and shellfish meat are good sources of most of the B vitamins. Flesh usually contains more than half the total amount of vitamin present in the fish. Fish roe is also a good source of these vitamins.

5.1. FAT SOLUBLE VITAMINS

5.1.1. Vitamin A and carotenoids

Vitamin A, as retinoids (primarily retinyl esters), is abundant in some animal-derived foods, whereas carotenoids are abundant in plant foods as pigments. They are available in high amounts in fish liver oils and small amounts in fish muscles, and hence fish is not a major source of this vitamin. Vitamin A is vulnerable to oxidation and overcooking. The concentration is expressed as 1 IU = 0.3 μg of all the trans-retinol or 0.6μg of β-carotene. Reversed-phase HPLC followed by UV detection is the most common method of analysis, for retinoids and carotenoids, and is given in AOAC official methods 2001.13 and 2005.07 (AOAC, 2006).

5.1.1.1. Vitamin A (retinol) estimation by HPLC method

The sample saponified in basic ethanol-water solution is neutralized to convert fats to fatty acids and retinol esters to retinol to be quantified in HPLC attached with UV detection at 313 or 328 nm.

Apparatus

1. HPLC system with UV detector
2. Column -RP C18, 10 m (4.6 mm i.d)
3. Flow rate - 1.0 to 2.0 ml/min
4. Injection volume – 20 µl
5. Detector wavelength – 313 or 328 nm

Reagents

1. Glacial acetic acid
2. Methanol
3. Ethanol, 95%
4. Tetrahydrofuran
5. Hexane
6. Pyrogallic acid, crystals
7. **Mobile phase:** Prepare a mobile phase by mixing 860 ml methanol and 140 ml of water and degass it either by stirring overnight or by mechanical degasser
8. THF-ethanol, 50:50
9. **Potassium hydroxide solution, 50%**

   Vitamin A working standard (15µg/ml): Weigh 50 mg of Vitamin A acetate concentrate (USP) – (equivalent to 30 mg of retinol/g of oil) in a 100 ml volumetric flask, add about < 3ml of acetone to dissolve and then dilute to volume with 95% ethanol. This solution is stable for 2 weeks at 4°C in dark.

   Retinyl palmitate standard: Weigh 55 mg of retinyl palmitate all-trans into a 100 ml of the volumetric flask, add 50 mg of pyrogallic acid, dissolve and then dilute to volume with hexane. Take 5 ml using a pipette to another 100 ml volumetric flask and dilute to volume with 95% alcohol. This solution is stable for 2 weeks at 4°C in dark.

Extraction and saponification

Weigh accurately 5±0.5 g of a dry or wet fish sample in a 125 ml conical flask. Add 3 ml of water, 40 ml of 95% ethanol, and 50 mg of pyrogallic acid, along with few glass beads to promote boiling. Then, add 10 ml of 50% KOH solution to the flask and reflux for 45 min. Cool the content and add 10 ml of glacial acetic acid to neutralize the KOH and cool again. Transfer the content to 100 ml volumetric flask and dilute to volume with THF-ethanol (50:50). Leave 1 h at room temperature or overnight at 5°C to precipitate fatty acid salts.

Preparation of vitamin standards

High standard: Transfer 5 ml of vitamin A working standard into a 125ml volumetric flask and add 25 ml of 95% ethanol. Continue the procedure with the addition of pyrogallic acid.

Intermediate standard: Transfer 2 ml of vitamin A working standard into another 125ml volumetric flask, and add 33 ml of 95% ethanol. Continue the procedure with the addition of pyrogallic acid.

Low standard: Transfer 0.5 ml of vitamin A working standard into another 125ml volumetric flask, and add 37.5 ml of 95% ethanol. Continue the procedure with the addition of pyrogallic acid.

HPLC analysis

Switch 'ON' the HPLC system, warm-up, and equilibrate for 30 min with the mobile phase at a flow rate of 1 ml/min. Then, inject 20 µl of vitamin A standard after saponification, and adjust the mobile phase to achieve a resolution of 1.5 or better for cis and transforms. All-trans retinol elutes in 9 min and cis retinol as a small peak just before all transforms. Then, inject 20 µl of high, medium, and low standards. Adjust the detector sensitivity to give peak heights of 50-90% of full scale for the high vitamin standard. Then, inject 20 µl of the sample solution. If retinol in test exceeds the peak height of the high standard, dilute the sample using 10 ml of 50% KOH, 40 ml of 95% ethanol, 10 ml of glacial acetic acid and 40 ml of THF-95% ethanol. Calculate the vitamin concentration by comparison of peak areas or peak height of samples with those of standards.

Calculation

Calculate relative factors for low, medium and high standards as follows:

$$\text{Response factor for vitamin A using USP standard} = \frac{Qs \times Vs \times Cs}{Ps \times 10,000}$$

where, Ps – peak area of the standard, Vs – volume of working standard, ml; Cs- concentration of working standard, (mg/g as retinol); Qs – weight of standard taken for reaction, mg; 10,000 – dilution factor for vitamin A standard

$$\text{Response factor for vitamin A using retinyl palmitate} = \frac{Qs \times Vs \times PUs \times 0.5458}{Ps \times 200}$$

where, PUs – purity % given by supplier divided by 100; Qs – weight of retinyl palmitate, mg; Ps - peak area of the standard, Vs – volume of working standard, ml; 0.5458 – ratio of retinol to retinyl palmitate molecular weights; 200 – combined dilution factor/ conversion from mg to µg.

Use the average relative factor for sample quantification. Measure the peak areas to retinol in the sample extracts. Measure the cis 13 peaks. Multiply the peak area of the cis 13 retinol peak by 1.08 (to compensate for the difference in absorbance compared to the trans-isomer). Add the corrected peak for the 13 cis-isomer to that of all-trans isomer to give the peak area of the sample solution. Calculate the vitamin A using the following equation.

$$\text{Vitamin A (µg/g as retinol)} = \frac{RFs \times P \text{ sample} \times 100}{W}$$

where, RF – response factor for vitamin A; Psample – total test sample peak area of all-trans and 13 cis retinol, 100 – dilution volume of sample, ml; weight of the sample, g

5.1.2. Vitamin D

Vitamin D in foods belongs to two major physiological forms: vitamin D2 and vitamin D3. In animal foods, vitamin D available as vitamin D3 and 25-hydroxy vitamin D. Fish liver is an excellent source of vitamin D. Finfish and shellfish are also natural contributors. Food processing, cooking, and storage of foods do not generally affect their concentration. The concentration is expressed in 1 IU = 0.005 µg of 25-hydroxy vitamin D or 0.025 µg of vitamin D2 or vitamin D3. HPLC method using an UV absorbance detector used quantify vitamin D and protocol is given in the AOAC official method 995.05 (AOAC, 2006).

5.1.2.1. Vitamin D estimation by HPLC method

Saponification and cleanup procedures first remove fats. After saponification, extract unsaponified lipids, including vitamin D with diethyl ether: petroleum ether (1:1), before the determination by reversed-phase HPLC, equipped with a UV detector at 265 nm. HPLC can separate vitamin D2 from vitamin D3.

Apparatus

1. HPLC system with UV detector fitted with column: C18, 5 mm particle size, 4.6 mm id
2. Solid-phase extraction (SPE) column - silica; 500 mg/2.8 ml.
3. Vacuum manifold for SPE column

Reagents

1. HPLC grade - *n*-hexane, dichloromethane, acetonitrile, isopropanol, methanol, ethyl acetate.
2. Ethanol, 95%.
3. Phenolphthalein solution, 1% in ethanol
4. **Dichloro methane–isopropanol (IPA)**

   DCM:IPA 99.8: 0.2 ratio (v/v) – Transfer 2 ml of isopropanol into 1 L volumetric flask and dilute to volume with dichloromethane.

   DCM: IPA 80 : 20 ratio (v/v) – Transfer 200 ml of isopropanol into 1 L volumetric flask and dilute to volume with dichloromethane and mix.
5. Acetic acid, 10%.
6. **Ethanolic potassium hydroxide (KOH)**

   Dissolve 140 g KOH in 310 ml of absolute ethanol and add 50 ml of distilled water. This solution is prepared fresh before use.
7. Mobile phase - Gradient mixture of acetonitrile, methanol, and ethyl acetate.
8. **Vitamin D2 standard solutions**

   Stock (180 mg/ml): Weigh accurately 45 mg vitamin D2 in a 250 ml volumetric flask and dilute to volume with absolute ethanol

   Working (2.88 mg/ml): Transfer 4.0 ml of the stock solution into 250 ml of a volumetric flask, and dilute to volume with absolute ethanol. This solution is stable for 7 days at 4°C.

Internal standard solution (46 ng/ml) – Use this standard for quantitative analysis of vitamin D3. Transfer 4.0 ml of working standard solution into 250 ml of a volumetric flask, and dilute to volume with absolute ethanol. This solution is stable for 7 days at 4°C.

9. **Vitamin D3 standard solutions**

Stock (180 mg/ml) – Weigh accurately 45 mg vitamin D3 in a 250 ml volumetric flask, and dilute to volume with absolute ethanol

Working (2.88 mg/ml): Transfer 4.0 ml of the stock standard solution into 250 ml of the volumetric flask, and dilute to volume with absolute ethanol. This solution is stable for 7 days at 4°C.

Internal standard solution (46 ng/ml) – Use this standard for quantitative analysis of vitamin D2. Transfer 4.0 ml of working standard solution into 250 ml of volumetric flask, and dilute to volume with absolute ethanol. This solution is stable for 7 days at 4°C.

Procedure

Transfer 4.0 ml of vitamin D2 internal standard solution and 4.0 ml of vitamin D3 internal solution into a conical flask and add 15.0 ml of distilled water. Prepare the test sample such that the target vitamin D concentration is 0.5 IU/ml. Transfer 15.0 ml to another conical flask and add 4.0 ml of internal standard solution. Use the vitamin D2 standard solution to quantify vitamin D3 and use vitamin D3 standard solution to quantify vitamin D2.

Saponification and Extraction

To the standard and test sample, add 15.0 ml of ethanolic KOH solution, close the flasks with stoppers, and place in water bath shaker for 30 min at 60°C. Cool and transfer the contents to 250 ml of separating flask with rinsing. To extract, add 60 ml hexane to the flask, shake vigorously for 90 sec and allow the layers to separate for 10 min. Drain the aqueous layer and discard it. Then, add 15.0 ml of distilled water to the hexane layer remaining in a flask, shake vigorously and allow the layers to separate. Discard the aqueous layer again. Then, add a drop of phenolphthalein solution and 15.0 ml of distilled water, followed by 10% acetic acid solution dropwise with shaking until washing is neutral to phenolphthalein (colorless). Drain the aqueous layer and discard it. Finally, drain the hexane layer through sodium sulfite is taken on a filter paper to a 100 ml round bottom flask.

Evaporation and Solid-Phase Extraction

Place the round-bottom flask on a rotary evaporator and evaporate the hexane to dryness at 40°C. Immediately after evaporation, add 2.0 ml of dichloromethane–IPA solution 1 to the flask. Prepare the SPE column and wash with 4.0 ml of dichloromethane–IPA solution 2 and 5.0 ml of dichloromethane–IPA solution 1. Transfer the solution from the round-bottom flask to the SPE column. Use 1.0 ml of dichloromethane–IPA solution 1 to wash the flask and transfer the rinsing to SPE column. Then, use 2.0 ml dichloromethane–IPA solution 1 to wash the SPE column and discard this fraction. Use 7.0 ml of dichloromethane–IPA solution 1 to elute vitamins D2 and D3 into a disposable culture tube. Place the culture tube in a water bath and evaporate the dichloromethane–IPA solution at 40°C using nitrogen. Reconstitute with 1.0 ml of acetonitrile and use for HPLC analysis

Chromatographic Determination

Inject the standards mixture onto the HPLC column at beginning, middle, and end of each run of the test solution. The flow rates and concentration of mobile phase components are given below. Set the injection volume as 25 µl. Set the column temperature at 27°C. Set the UV detector wavelength at 265 nm. The retention time of vitamin D2 is 19.5 min and vitamin D3 is 23 min.

Flow rates and concentrations of mobile phase components

Time, min	Flow rate, ml/min	Acetonitrile,%	Methanol,%	Ethyl acetate,%
0.0	0.7	91.0	9.0	0.0
28.0	0.7	91.0	9.0	0.0
28.5	2.5	0.0	0.0	100.0
31.0	2.5	0.0	0.0	100.0
31.5	2.5	91.0	9.0	0.0
33.0	2.5	91.0	9.0	0.0
34.0	0.7	91.0	9.0	0.0

Use internal standards and peak heights to calculate the content of vitamins D2 and D3.

Calculation

1. Concentration of vitamin D_2 in standards mixture

$$CSD_2 \text{ (IU/ml)} = \frac{W \times 4 \times 4 \times 4 \times 4000}{250 \times 250 \times 250} \times 1.05$$

where, W = weight of vitamin D_2 standard, mg; 4 = dilution factor; 40 000 = IU vitamin D/mg; 250 = volumes of sub sequent dilutions of vitamin D_2 standard solutions; 1.05 = correction factor for pre-vitamin D

2. Concentration of vitamin D3 in standards mixture (CSD_3) as above using vitamin D_3 standard

3. Response ratio of vitamin D_2 in standards mixture

$$RSD_2 = \frac{PSD_2}{PSD_3}$$

where PSD_2 = peak height of vitamin D_2 in standards mixture; PSD_3 = peak height of vitamin D_3 in standards mixture.

4. Response ratio of vitamin D_3 in standards mixture (RSD_3) as above, using peak heights of vitamins D_3 and D_2, respectively.

5. Response ratio of vitamin D- in test portion (RTD_2)

$$RTD_2 = \frac{PSD_2}{PSD_3}$$

where PTD_2 = peak height of vitamin D_2 in test portion; PTD_3 = peak height of vitamin D_3 in test sample.

6. Response ratio of vitamin D_3 in test portion (RTD_3) as above, using peak heights of vitamins D_3 and D_2, respectively.

7. Concentration of vitamin D_2 (CTD_2) or D_3 (CTD_3) in test portion as follows:

$$CTD_2 \ (IU/ml) = \frac{RTD_2}{RSD_2} \times \frac{CSD_2}{V_t} \times U \times D$$

$$CTD_3 \ (IU/ml) = \frac{RTD_3}{RSD_3} \times \frac{CSD_3}{V_t} \times U \times D$$

Where, V_t = volume of test portion, ml; U = conversion factor to appropriate units; D = dilution factor for diluted powders or liquids.

5.1.3. Vitamin E

Vitamin E is the most effective fat-soluble vitamin. There are eight forms of natural vitamin E: four tocopherols (α, β, γ, and δ tocopherols), and four tocotrienols (α, β, γ and δ tocotrienols). Aquatic food, including shellfish, contains small amounts of vitamin E. Fatty fish have more vitamin E than lean fish. Tocopherol and antioxidants protect fish oils from rancidity. During processing, losses of vitamin E can occur quite rapidly due to oxygen, light, heat, and various metals (primarily iron and copper), or presence of free radicals, that initiate autoxidation. Vitamin E is expressed in mg of α tocopherol equivalents. There are colorimetric and polarimetric methods for the determination of vitamin E, as given in AOAC official methods 948.20, 971.80, and 975.45 (AOAC, 2006). The HPLC is the most preferred method for the determination of different tocopherols and tocotrienols, as given in the AOAC official method 992.03 (AOAC, 2006).

5.1.3.1. Vitamin E estimation by spectrophotometric method

Total unsaponifiable matter extracted and fractioned by TLC to recover the tocopherol fraction or their components, which then react with the mixture of ferric chloride, and 2, 2' dipyridyl (Emeric – Engel method) to form a pink color complex having absorbance at 520 nm.

Reagents

1. **Standard α- tocopherol**: Dissolve 200 mg of standard mixture with equal parts of α, β and δ tocopherol in 10 ml heptane. In the absence of standard tocopherols, prepare the solution of tocopherols from a mixture containing equal parts of soybean oil and sunflower oil. This solution gives a mixture in approximately equal parts of the three tocopherols.

2. Pyrogallol, 5% in ethanol. Prepare this solution immediately before use.

3. Potassium hydroxide,160 g / 100 ml. Prepare this solution immediately before use.

4. Diethyl ether

5. Phenolphthalein, 1% in ethanol

6. Ethanol, benzene, hexane, n – heptane

7. Developing solvent: Hexane: diethyl ether (70:30)

8. Ferric chloride solution, 0.2% in ethanol. Prepare immediately before use

9. 2, 2' dipyridyl solution: 0.2 % in ethanol

10. Spray reagent: Ferric chloride and 2, 2' dipyridyl solution (1:1)

11. Elution solvent: Heptane and absolute ethanol (2:1)

Separation of tocopherol fraction

Weigh accurately about 1 g of fat or oil in a 50 ml round bottom flask and add 4 ml of pyrogallol ethanolic solution to it. Attach the flask to the air-condenser.

Boil the content, and when boiling starts, add 1 ml of KOH solution and boil for 3 min. Cool the flask and add 25 ml of distilled water. Transfer the content to a separating funnel and rinse the flask with 40 ml of diethyl ether. Make the first extraction and remove the diethyl ether layer. Make the second extraction with 25 ml of diethyl ether and combine the extracts. Wash the combined extract with 20 ml of distilled water with vigorous shaking until the washing liquid does not turn pink on the addition of a drop of phenolphthalein solution. Transfer the ether layer into a 100 ml flask and evaporate by distillation using a rotary evaporator or boiling water bath. Then, add 1 ml of ethanol and 4 ml of benzene to the dry residue and evaporate to dryness under the stream of nitrogen. Re-dissolve the residue in 1 ml of hexane and transfer to a 15 ml screw-cap test tube using a minimum hexane rinsing. Evaporate the hexane under a vacuum or a stream of nitrogen. Then, add exactly 1 ml of heptane with the aid of a volumetric pipette.

Thin-layer chromatography

Carry out all operations in a dimly illuminated room. First, saturate the developing tank with developing solvent. Activate the silica gel G 20 x 20 TLC plate for 30 min at 110°C. Then, apply 10 µl of the reference solution of tocopherols at one spot. Then, in a band (2 cm wide), apply in small drops 50 µl of the unsaponifiable matter solution. Introduce the plate into a developing tank, fit the lid, and develop the plate with the mobile solvent until the solvent front reaches 1 cm below the top of the plate. Remove the plate after development, and spray with a mixture of equal volumes of ferric chloride and 2, 2' dipyridyl solution and place in the dark for 5-10 min. The appearance of deep pink spots indicates the location of tocopherols derived from the reference solution. Trace the plate for the presence of tocopherols with the help of the reference marks. Mark the spots about 2 cm wide, remove along with the silica, and transfer to an elution column. Use 5-6 ml of elution solvent to extract the silica and then evaporate the solvent under vacuum or stream of nitrogen.

Spectrophotometric determination

Carry out all the operations in a dark room. To the tocopherol fraction, add 3.6 ml of ethanol, 0.2 ml of 2, 2' dipyridyl solution, and 0.2 ml of ferric chloride solution, mix well, and allow to stand for 10 min. Make a blank without tocopherol but with the same quantities of reagents. Measure the absorbance of ethanol blank at 520 nm. The blank must give an absorbance less than 0.05.

Calculation

$$\text{Each tocopherol (mg / 100 g)} = \frac{(A-A_o) \times F \times V \times 100}{v \times m}$$

where, A - absorbance of test solution; A_o - absorbance of blank solution; V - volume of heptane, ml; v - volume of the unsaponifiables applied to the TLC plate, μl; m – mass of the fatty material, g; F - Spectrophotometric factor different for each tocopherol (98 - α tocopherol; 90 - β and δ tocopherol; 75 - γ tocopherol)

5.1.4. Vitamin K

There are two forms of vitamin K in nature: phylloquinone and menaquinone. In most fish and shellfish, there are only minute amounts of vitamin K. Vitamin K is quite stable to oxidation in most food processing. It is unstable to light, alkali, strong acid, and reducing agents. Vitamin K is expressed as mg of vitamin K per gram in food. The current methods used to determine vitamin K in foods include HPLC procedure using fluorescence or electrochemical detection systems. AOAC official methods 992.27 and 999.15 are available for the determination of phylloquinone and menaquinone in infant formulas (AOAC, 2006). Vitamin K is extracted from foods with organic solvents such as ethanol, isopropanol, and acetonitrile and purified by solid-phase extraction with silica cartridges before the resolution by HPLC.

5.2. WATER SOLUBLE VITAMINS

5.2.1. Vitamin B complex

5.2.1.1. Thiamin (Vitamin B1)

Thiamin, vitamin B1 exists in interconvertible phosphorylated forms, as thiamin monophosphate, thiamin pyrophosphate, and thiamin triphosphate. Most fish and shellfish have small amounts of thiamin. Thiamin is most stable between pH 2-4 and gets oxidized at alkaline pH. Thiamin is the most heat-labile vitamin. Raw fish and shellfish contain thiaminase, an enzyme that destroys thiamin. Sulfhydryl groups and other reducing agents protect thiamin from thiaminase reaction. Thiaminase gets inactivated during cooking process. Thiamin is expressed as mg of thiamin in foods. The method used for thiamine estimation are fluorometric AOAC official methods, 942.23, 953.17, 957.17, and 986.27 (AOAC, 2006), microbiological assay, HPLC methods, and GC procedures.

5.2.1.1.1. Thiamin estimation by fluorometric method

Alkaline potassium ferricyanide oxidizes thiamine to thiochrome, a fluorescent compound, which is extracted in isobutyl alcohol and measured in a spectrofluorometer.

Reagents

1. NaOH, 15%
2. Potassium ferricyanide, 1%
3. Isobutyl alcohol
4. H_2SO_4, 0.1N
5. **Standard thiamine HCl**

   Stock (100µg/ml): Dissolve 50 mg of thiamine HCl in 500 ml of 0.1N H_2SO_4 containing 25% alcohol.

   Working I (5µg/ml): Dilute 5 ml of the stock standard solution with 100 ml of 0.1 N H_2SO_4.

   Working II (0.25µg/ml): Dilute 5 ml of the working standard I solution with 100 ml of 0.1N H_2SO_4.

Procedure

Weigh accurately, 5±0.5 g of a dry sample in a stopper flask, add 100 ml of 0.1 N H_2SO_4 and allow to stand overnight. Filter the content and discard the first 10-15 ml of the filtrate. Take 10 ml of the extract in a separating funnel. Also, take 10 ml of the working standard in another separating funnel. To extract, add 3 ml of 15% NaOH into the separating funnel followed by four drops of ferricyanide solution. Shake the content gently for 30 sec and add 15 ml of isobutanol rapidly. Shake the content vigorously for 60 sec and allow the layers to separate. Drain off the bottom layer carefully and discard it. Then add a spatula of sodium sulfate directly into the separately funnel and swirl gently to clarify the extract. Collect the clear extract from the top using Pasteur pipette into a clean dry test tube.

To prepare a sample blank, take 10 ml of the extract and follow the above procedure except the addition of ferricyanide. To prepare a standard blank, take 10 ml of the working standard and follow the above procedure, except the addition of ferricyanide. Set the spectrofluorometer at 366 nm excitation wavelength and 425 nm emission wavelength, and read the fluorescent intensities of standard, standard blank, sample, and sample blank.

Calculation

$$\text{Thiamine } (\mu g/100g) = \frac{0.25 \times 10}{a - a^1} \times \frac{(x - x^1) \times 100}{10} \times \frac{10}{5}$$

where, a – fluorescence reading of standard; a^1 – fluorescence reading of standard blank; x – fluorescence reading of sample; x^1 – fluorescence reading of sample blank

5.2.1.2. Riboflavin (Vitamin B2)

Riboflavin, vitamin B2, acts as an integral component of two coenzymes: adenine dinucleotide (FAD) and flavin mononucleotide (FMN). It occurs naturally in foods as free riboflavin and as the protein-bound coenzymes, FAD and FMN. Seafood is generally a modest source of riboflavin. Some species like mackerel and squid are good to excellent sources. Fish consumed whole (e.g. smelt and sardines) are also good sources of riboflavin. Riboflavin is stable to heat, acid, oxidation, and during heat processing and cooking. Riboflavin expressed as mg of riboflavin in foods. Estimation of riboflavin is by fluorimetric AOAC official methods 970.65 and 981.15 (AOAC, 2006), microbiological assays, and HPLC methodologies using fluorescence detection. The HPLC method separates individual free riboflavin, FAD, and FMN in foods.

5.2.1.2.1. Riboflavin estimation by fluorometric method

Riboflavin fluoresces at wavelengths between 440-500 nm. The fluorescence intensity is proportional to the concentration of riboflavin in the solutions. The riboflavin content is measured in terms of the differences in fluorescence before and after the chemical reduction.

Reagents

1. H_2SO_4, 0.1 N
2. Sodium acetate, 2.5 M
3. Potassium permanganate, 4%
4. Hydrogen peroxide, 3%
5. Acetic acid, 0.02 M
6. Sodium hydrosulfite

7. **Riboflavin standard solution**

Stock (25 µg/ml): Dissolve 50 mg of riboflavin in 1.5L of distilled water and add 2.4 ml glacial acetic acid. Warm the solution cool and make up to 2 L volume with distilled water.

Working I (10 µg/ml): Dilute 20 ml of stock standard solution to 50 ml with distilled water.

Working II (1 µg/ml): Dilute 10 ml of working standard I solution to 100 ml with distilled water.

Procedure

Weigh accurately 5±0.5 g of a dry sample into a conical flask, add 50 ml of 0.1 N H_2SO_4 and mix well. Autoclave the mixture at 15 psi for 30 min or immerse in a boiling water bath for 30 min. Shake the content in the flask for every 5 min and allow to cool. Then, add 5ml of 2.5M sodium acetate, mix well, and allow to stand 1 h at least. Adjust the pH of the mixture to approximately pH 4.5 with acid or alkali, dilute to 100 ml with distilled water, and filter. Take 10 ml of the sample extract in 4 test tubes (two tubes for standard+sample and two tubes for sample). To the standard+sample tubes, add 1 ml of working standard II and mix well. To the sample tubes, add 1 ml of distilled water and mix well. To all the test tubes, add 1 ml of acetic acid, 0.5 ml of 4% $KMnO_4$, mix well, and allow to stand for 2 min. If oxidizable material is high, add another 0.5 ml of $KMnO_4$. Then, add 0.5 ml of 3% hydrogen peroxide to all the tubes. Permanganate color must destroy within 10 sec. Set the spectrofluorometer at an excitation wavelength of 440 nm and an emission wavelength of 565 nm. Measure the fluorescence (X) intensity of test solution containing 1 ml added standard riboflavin. Measure the fluorescence (B) intensity of the test solution containing 1 ml of added water. Then, add 20 mg of sodium hydrosulfite to the test solution containing 1 ml added standard riboflavin and measure the fluorescence (C) within 5 sec.

Calculation

$$\text{Riboflavin } (\mu g/100g) = \frac{B-C}{X-B} \times \frac{R}{S} \times \frac{V}{V_1} \times 100$$

where, X – fluorescence reading of the sample + riboflavin standard; B – fluorescence reading of the sample + water; C – fluorescence reading after addition of sodium hydrosulphite; R - standard riboflavin added $\mu g/V_1$ of sample solution; V - original volume of sample solution, ml; V_1- volume of sample solution taken for measurement, 10 ml; S – weight of fish, g

5.2.1.3. Niacin (Vitamin B3)

Niacin is also known as vitamin B3. The term 'niacin' is the generic descriptor for nicotinic acid and nicotinamide. These compounds are essential for the formation of the coenzymes, nicotinamide adenine dinucleotide (NAD) and nicotinamide adenine dinucleotide phosphate (NADP) in the body. Niacin is biosynthesized from the amino acid tryptophan by the body. Niacin contents in seafood considerably vary depending on their variety. Lean white fish and shellfish contain smaller amounts of niacin, whereas mackerel, salmon, and tuna are rich sources. Nicotinic acid is found mainly in plant and animal foods, contain nicotinamide. Niacin is not affected by thermal processing, light, oxygen, and pH. Niacin content is expressed as mg of niacin equivalents (NE) in food. Estimate states 60 mg of dietary tryptophan is equivalent to 1 mg of niacin. So, 1 mg NE is equivalent to 1 mg of niacin or 60 mg of dietary tryptophan. Niacin is determined by colorimetric AOAC official methods 961.14 and 975.41 (AOAC, 2006) using Konig reaction, microbiological assays, and HPLC methods.

5.2.1.3.1. Niacin estimation by spectrophotometric method

Nicotinic acid and nicotinamide reacts with cyanogen bromide to give a pyridinium compound which undergoes rearrangement yielding derivatives that couple with aromatic amine, and sulfanilic acid, giving rise to a colored compound having an absorption maximum at 420 nm.

Reagents

1. **Nicotinic acid standard**

   Stock (100µg/ml): Dissolve 50 mg of USP nicotinic acid standard in little alcohol and make up to 500 ml with distilled water. Store the solution at 10°C in a dark bottle.

   Working (4µg/ml): Dilute 2 ml of the stock standard in 50 ml of distilled water

2. **Sulphanilic acid solution, 10%** : Dissolve 20 g of sulphanilic acid in 170 ml of distilled water with the addition of ammonium hydroxide solution. Adjust the pH to 4.5 with dilute HCl (1:1) using bromocresol green as an indicator. Make up the volume to 200 ml with distilled water.

3. **Sulphanilic acid solution, 55%:** Dissolve 55 g of sulphanilic acid with 27 ml of distilled water and ammonium hydroxide solution. Adjust the pH to 7.0 with few drops of ammonium hydroxide/ dilute HCl acid. Make up the volume to 100 ml with distilled water

4. Cyanogen bromide solution, 10%

5. **Dilute NH$_2$OH solution:** Dilute 5 ml of 28% NH$_2$OH in 250 ml of distilled water

6. Dilute HCl/HBr (1:5)

7. **Phosphate buffer solution :** Dissolve 60 g of disodium hydrogen phosphate, Na$_2$HPO$_4$.7H$_2$O and 10 g of KH$_2$PO$_4$ in 200 ml of distilled water.

Procedure

Weigh accurately 40±0.5g of a tissue sample in a 1 L conical flask, add 200 ml of (NH$_2$)$_4$SO$_4$, and autoclave for 30 min at 120°C. Cool the mixture and adjust the pH to 4.5 with a standard NaOH solution using bromocresol green as an external indicator. Dilute the solution to 250 ml with distilled water and filter. From that, take 40 ml of the sample solution in a 50 ml volumetric flask, add 17 g of (NH$_2$)$_4$SO$_4$, and make up the volume with distilled water. Shake the content and filter to get the sample solution for color development. Similarly, take 40 ml of working standard in a 50 ml volumetric flask, add 17 g of (NH$_2$)$_4$SO$_4$ and makeup to volume with distilled water (1ml = 3.2µg of niacin).

Take 1.0 ml of the standard solution in two test tubes to serve as standard blank (A) and standard solution (B). To which, add 5.0 ml of distilled water to tube A. To all the test tubes, add 0.5 ml of dilute NH$_2$OH. Then, add 5.0 ml of cyanogen bromide solution to tube B. Finally, add 2.0 ml of 10% sulphanilic acid and 0.5 ml of dilute HCl/HBr to all the test tubes. Read the absorbance at 420 nm in a spectrophotometer within 30 sec.

Then, take 1.0 ml of the sample solution in two test tubes to serve as sample blank (C) and unknown sample (D). Then, add 5.0 ml of distilled water to tube C. To all the test tubes, add 0.5 ml of dilute NH$_2$OH. Add 5.0 ml of cyanogen bromide solution only to tube D. Finally, add 2.0 ml of 10% sulphanilic acid and 0.5 ml of dilute HCl/HBr to all the test tubes. Read the absorbance at 420 nm in a spectrophotometer with 30 sec.

Calculation

$$\text{Niacin (µg/g)} = \frac{3.2 \times B \times C}{A \times W}$$

where, B – absorbance of 1 ml of sample solution, C – total volume of sample extract, ml; A- absorbance of 3.2 mg of niacin; W – weight of sample, g

5.2.1.4. Pyridoxine (Vitamin B6)

Vitamin B6 is a water-soluble vitamin consisting of derivatives of 3- hydroxy-2-methylpyridine: pyridoxal (PL), pyridoxine (PN), pyridoxamine (PM), and their respective 5'-phosphates (PLP, PNP, and PMP). PLP is the metabolically active B6 vitamin. Fish are rich sources of vitamin B6. PLP bound to the apoenzyme by a Schiff base is the major form of vitamin B6 in fish. Vitamin B6 is unstable to light, stable in acidic conditions, heat processing, and storage. Pyridoxine is more stable to heat than pyridoxal and pyridoxamine. Vitamin B6 is expressed as mg of vitamin B6 in food. Microbiological, enzymatic, fluorometric, GC, and HPLC assays are available to measure vitamin B6. Currently, microbiological official method 961.15 (AOAC, 2006) and HPLC methods widely used to measure total vitamin B6.

5.2.1.4.1. Pyridoxine estimation by microbiological assay

Saccharomyces carlsbergenes requires specified vitamins for growth. Using a basal medium complete in all aspects except for pyridoxine and the growth responses of the organism compared quantitatively in a standard and unknown solution to determine vitamin B6.

Reagents

1. **Glucose solution**: Dissolve 100 g of glucose in 600 ml of distilled water.

2. **Stock solutions**

   Salts: Dissolve 1.1 g of KH_2PO_4, 0.85g of KCl, 0.25 g of $CaCl_2$, 0.25g of $MgSO_4$, 5 mg of $MnSO_4$, and 5 mg of $FeCl_3$ in 100 ml of distilled water. Add a few drops of HCl to dissolute the precipitate.

   Casein hydrolysate: Stir 100 g of vitamin free casein or its hydrolysate with 250 ml of 95% ethanol for 15 min in a beaker and filter. Repeat the extraction with fresh 250 ml of ethanol. Transfer the washed casein into a 1L round bottom two necked flasks and mix with 500 ml of constant boiling HCl (1:1). Fit the flask with a glass stopper, place it in a condenser and reflux over low heat for 12 h. Cool the reactant with a wet towel in the event of a vigorous reaction. When the hydrolysate gets concentrated, almost the entire HCl gets removed under reduced pressure. Introduce air in the flask to reduce bumping during the final stages through a bleeder tube placed well into the bottom of the flask. Set the distillation temperature at 75°C and complete the distillation by pressure reduction by a steam aspirator. Concentrate the hydrolysate until it becomes pasty and re-dissolve in 200 ml of water. Repeat the concentration process to remove excess HCl. Re-dissolve the hydrolysate paste in 700 ml of water, and adjust the pH to 3.5 with 40% NaOH. Add 20 g of activated charcoal to decolorize the solution within 5 min or sometimes 1h. Stir the content until the test filtrate is light straw. This step removes the leftover niacin in the alcohol washed casein. Filter the content through a fluted filter paper and adjust the pH to 6.8. Dilute the filtrate to 1L and store under toluene and over chloroform in a refrigerator. This solution forms a precipitate consisting mainly of tyrosine. Shake the solution before use.

3. *Biotin solution: Dissolve* 4 mg of biotin in 100 ml of distilled water by warming. Dilute 10 ml of this stock solution to 100 ml of distilled water. Prepare this solution fresh before use.

4. *Calcium pantothenate solution*: Dissolve 5 mg of calcium D- pantothenate, 50 mg of meso-inositol, and 5 mg of niacin in 100 ml of distilled water by warming

5. **Thiamin solution:** Dissolve 5 mg of thiamine HCl in 25 ml of distilled water

6. **Potassium citrate buffer:** Dissolve 10 g of potassium citrate and 2 g of citric acid in 100 ml of distilled water.

7. **Basal medium**

   Mix solution 1 – 10ml, solution 2 – 8 ml, solution 3 – 0.4 ml, solution 4 – 10 ml, solution 5 – 0.25 ml, and solution 6 – 10ml, adjust the pH to 5.5 and makeup the volume to 100 ml. Finally, add 60 ml of glucose solution to the above solution.

8. **Stock culture medium**

Malt extract	- 0.3 g
Glucose	- 1.0 g
Yeast extract	- 0.3 g
Peptone	- 2.1 g
Agar	- 1.5 g

   Dissolve all the above ingredients by heating in 100 ml of distilled water. Dispense 8 ml portion into test tubes and sterilize at 121°C for 15 min to prepare slants. Sub-culture *Saccharomyces carlsbergensis* every fortnight in slant culture.

9. **Pyridoxine standard**

   Stock (500μg/ml): Dissolve 50 mg of pyridoxine in 100 ml of distilled water

   Working I (50μg/ml): Dilute 1 ml of the stock to 10 ml of distilled water.

   Working II (5μg/ml): Dilute 1 ml of the working I to 10 ml of distilled water.

Procedure

Preparation of inoculum

Wash the freshly grown agar slope after 20 h of incubation with sterile saline and transfer into a sterile tube. The culture turbidity of 15% or 85% transmission corresponds to 0.071 OD at 600 nm.

Preparation of standards

Transfer 0, 5, 10, 15, 20, 25, 30, 40, and 50 µg of pyridoxine standards into a series of volumetric flasks and make up the volume with sterile distilled water. Prepare the standards daily.

Preparation of sample

Weigh accurately 2±0.2 g of the fish tissue sample, suspend in 20 ml of water and 20 ml of 0.5 N H_2SO_4, and autoclave at 121°C. Adjust the pH to 5.5 and make up the volume up to 100 ml with distilled water. Mix the content well, filter, and take suitable aliquots in triplicate for pyridoxine assay at 3 concentrations.

Spectrophotometric determination

Transfer the standards or samples into a 5 ml conical flask and make up the volume up to 1.0 ml with distilled water. Then, add 8 ml of casein hydrolysate and sterilize at 121°C for 12 min. To the medium, add 1 ml of inoculum under the sterile condition and incubate it for 20 h at 37°C. Measure the turbidity in a spectrophotometer at 600 nm using the un-inoculated blank tube to set the instrument at zero.

Calculation

Draw a standard curve by plotting turbidity reading (OD) on the X-axis against the concentration of the vitamin on the Y-axis. Determine the vitamin content in the sample by interpolation of the reading on the standard curve. Determine the average vitamin content for 1 ml of the test solution from values obtained from not less than 3 tubes.

$$\text{Pyridoxine (mg/g)} = \frac{\text{Average pyridoxine (mg/ml)} \times \text{dilution factor}}{\text{Weight of the sample, g}}$$

5.2.1.5. Vitamin B12

Vitamin B12 is a water-soluble vitamin and a family of compounds called cobalamins. Cobalamins contain cyanocobalamin, hydroxocobalamin, and two coenzyme forms: 5'-deoxy adenosylcobalamin and methylcobalamin. Vitamin B12 is synthesized by bacteria growing in soil and water. Fish and shellfish are rich sources of vitamin B12, mainly salmon, herring, sardines, mackerel, oyster and clam. The most prevalent forms of vitamin B12 in seafood are adenosylcobalamin, hydroxocobalamin, and methylcobalamin. Canned fish contains sulfitocobalamin. Vitamin B12 is generally stable to thermal processing. Vitamin B12 is expressed in mg or µg of vitamin B12 in food. The microbiological assay is widely used for the determination of total vitamin B12 in foods.

5.2.2. Vitamin C

Vitamin C is a water-soluble vitamin that occurs in two forms: reduced ascorbic and oxidized dehydroascorbic acid. Ascorbic acid reversibly oxidizes to dehydroascorbic acid first and further to inactive and irreversible compound, diketoglutamic acid. There are two enantiomeric pairs: L- and D-ascorbic acid, and L-and D-isoascorbic acid. L-ascorbic acid and D-iso ascorbic acid have the biological activity of vitamin C. L-ascorbic acid and dehydroascorbic acid are naturally occurring forms of vitamin C. D-iso ascorbic acid is synthesized commercially and used as an antioxidant in foods, usually processed and canned fish. Fish and seafood are not considered as good sources of vitamin C. Vitamin C susceptible to oxidation during the processing, storage, and cooking of foods, especially under alkaline conditions. Vitamin C contents commonly expressed in mg of vitamin C in food. Methods for determination of total vitamin C in foods include the titrimetric method using oxidation-reduction indicators (AOAC, 2006), fluorometric method, enzymatic methods, and HPLC methods using UV, fluorescence, or electrochemical detection. HPLC method can simultaneously separate L-ascorbic, dehydroascorbic, and D-iso ascorbic acids.

5.2.2.1. Vitamin C estimation by titrimetric method

The titrimetric method uses 2,6-dichlorophenol in measuring ascorbic acid, not dehydroascorbic acid. This method not used for processed food products containing D-iso ascorbic acid.

Ascorbic acid reduces 2,6-dichlorophenol indophenol dye to a colorless leuco-base. The ascorbic acid gets oxidized to dehydroascorbic acid to form a blue-colored compound, however, the endpoint is the appearance of pink color in acid medium, which can be quantified by titration.

Reagents

1. Oxalic acid, 4%.

2. Dye Solution

Dissolve 42 mg sodium bicarbonate in a small volume of distilled water, add 52 mg of 2, 6-dichloro phenol indophenol and make up the volume to 200 ml with distilled water

3. Ascorbic acid standard solution

Stock (1mg/ml): Dissolve 100 mg L-ascorbic acid in 100mL of 4% oxalic acid solution.

Working (100mg/ml): Dilute 10ml of the stock solution to 100 ml with 4% oxalic acid.

Procedure

Transfer 5ml of the working standard solution into a conical flask and add 10ml of 4% oxalic acid. Titrate the content against the dye (V1, ml) until the appearance of a pink color, which persists for a few mins. The amount of the dye consumed is equivalent to the amount of ascorbic acid. Extract the sample (0.5 - 5g) with 10 ml of 4% oxalic acid and make up the volume up to 100 ml with distilled water. Centrifuge the content at 5000 ppm at 15 min and filter. Then, take 5 ml of the filtrate, add 10 ml of 4% oxalic acid and titrate against the dye (V2, ml), until the appearance of pink color, which persists for a few mins.

Calculation

$$\text{Ascorbic acid (mg/100g)} = \frac{0.5 \text{ mg}}{V_1 l} \times \frac{V_2 l}{5 mL} \times \frac{100 mL}{\text{Wt. of the sample}} \times 100$$

CHAPTER 6

ANALYSIS OF MINERALS

Fish and shellfish are considered as a source of essential minerals and a well-balanced supply of minerals. Fish is a valuable source of calcium and phosphorus, and also iron, copper and selenium. Marine fish has a high content of iodine. Sodium content is relatively low in fish and hence suitable as a low-sodium diet.

Ash content is the total amount of minerals present within a food, whereas mineral content is the amount of specific inorganic components, such as Ca, Na, K, and Cl. The total ash content never exceeds 1-2% of the edible portion. Analyses of ash and mineral content of foods are for many reasons like nutritional labeling, fish quality, microbiological quality, fish nutrition, and fish processing.

6.1. TOTAL ASH

Ash is the inorganic residue remaining after the removal of water and organic matter by heating in the presence of oxidizing agents. It provides a measure of the total amount of minerals within a food. Two main types of analytical procedures used are based on principles like dry ashing and wet ashing. Ashing is also used as the first step in the preparation of samples for analysis of specific minerals by atomic spectroscopy or traditional methods.

6.1.1. Dry ashing

Dry ashing uses a high-temperature muffle furnace maintained at temperatures between 500 and 600°C. At high temperatures, all the organic substances get burned to CO_2, H_2O, and N_2 in the presence of the oxygen in the air, while water and other volatile materials get vaporized. Most minerals get converted to oxides, sulfates, phosphates, chlorides, or silicates. Most minerals are less volatile at high temperatures, and some are volatile that partially disappear, *e.g.,* iron, lead, and mercury. It is advisable to use an alternative ashing method using low temperatures. to determine the concentration of these substances. Crucibles made with porcelain used for ashing, as they are relatively inexpensive, tolerant to high temperatures (< 1200°C), resistant to acid, and easy to clean. AOAC Official Methods of Analysis provides the procedure for the determination of ash content (AOAC, 1995).

Procedure

Weigh accurately 2±0.2 g of the dried fish sample in a silica crucible and ash in a muffle furnace set at 500 – 600°C for 24 h until the residue is white. Cool the residue in a desiccator and weigh again. Express the ash content on a dry weight basis.

Calculation

$$\text{Ash, \% (dry weight basis)} = \frac{\text{Wt. of ash}}{\text{Wt. of fish}} \times 100$$

Microwave heating can dry ash samples. It removes most of the moisture at a relatively low heat and then converts the sample to ash at relatively high heat. The microwave instrument reduces the time of analysis to < 30 min.

6.1.2. Wet Ashing

Wet ashing is employed when a specific mineral analysis of the sample is required. Ashing removes the organic matrix surrounding the minerals by digestion. In this process, a dried sample placed in a flask is heated with strong acids and oxidizing agents such as nitric, perchloric, or sulfuric acid, until all the organic matter get digested, leaving the mineral oxides in solution. The time and temperature required for the digestion depend on the type of acids and oxidizing agents. Generally, it takes from 10 min to a few hours at temperatures of about 350°C for complete digestion.

6.1.3. Acid insoluble ash (sand content)

Ash is the inorganic material obtained after the removal of the organic matter by burning at an elevated temperature of around 500°C. The ash left out after solubilization with acid is known as acid-insoluble ash. It indicates the presence of sand.

Procedure

Weigh accurately 2±0.2 g of the dried sample in a silica crucible and ash in a muffle furnace at 500 – 550°C for 24h, until the residue becomes ash. Add 25 ml of 5N HCl to the ash in the silica dish, cover with a watch-glass, and heat on a water- bath for 10 min. Cool the contents of the dish and filter through Whatman filter paper No.42. Wash the filter with water until the washing is free of acid. Return the filter with the residue to the dish, and keep it in a hot air-oven set at 135±2°C for 3 h. Place the dish in a muffle furnace set at 600±20°C for 3 h, and ignite filter paper with the residue. Cool it in a desiccator, and weigh.

Calculation

$$\text{Acid insoluble ash, \% (dry wt. basis)} = \frac{(W_2 - W_0)}{(W_1 - W_0)} \times 100$$

$$\text{Acid insoluble ash, \% (wet wt. basis)} = \frac{(100 - M) \times \text{Ash \% (dry wt)}}{100}$$

where, W_0 - weight of empty dish, g; W_1 - weight of ash dish, g; W_2 - weight of acid insoluble ashdish, g; M - moisture content, %;

6.2. SPECIFIC MINERALS

Knowledge on the concentration and the type of specific minerals in aquatic food is of importance. The most effective means of determining the type and concentration of specific minerals is by atomic absorption or emission spectroscopy, often to ppm levels. Traditional methods of mineral analysis are being replaced by modern methods. Some minerals can also be measured using traditional methods (AOAC, 1995). Ashing can be carried out by anyone of the methods described above to isolate the minerals from the organic matrix before the analysis. There are several traditional methods for specific mineral analysis.

6.2.1. Gravimetric analysis

In this method, precipitate the element from solution by the addition of a reagent first that reacts with it to form an insoluble complex. Separate the precipitate from the solution by filtration. Rinse the precipitate, dry, and weigh. Determine the amount of mineral present in the original sample from the knowledge of the chemical formula of the precipitate. For instance, determine the amount of chloride in a solution by adding excess silver ions to form an insoluble silver chloride precipitate, because AgCl contains 24.74% Cl. Gravimetric methods are suitable for large food samples having relatively high concentrations of the mineral but not for trace element analysis.

6.2.2. Spectrophotometric methods

Spectrophotometric methods depend on the change in color of any reagent when it reacts with a specific mineral in solution. Quantify the color change by measuring the absorbance of the solution at a specific wavelength using a spectrophotometer. These methods can determine the concentration of different minerals.

6.2.2.1.Total Phosphorus

The spectrophotometric method is applicable for the determination of total phosphorus in food products and even the added phosphate levels in fish and shellfish (AOAC, 1975).

Vanadate is the colorimetric reagent for the determination of phosphorus. The phosphorus content is determined by the addition of vanadate-molybdate reagent to the ash solution, to form a colored complex (yellow-orange) with the phosphorous having an absorption maximum.

Reagents

1. Hydrochloric acid (HCl) (1:3)
2. Nitric acid (HNO_3)
3. Perchloric acid ($HClO_4$), 70%
4. **Molybdate solution**
   Dissolve 2 g of ammonium molybdate tetrahydrate [$(NH_4)_6Mo_7O_{24}.4H_2O$] in 50 ml hot distilled water and cool.
5. **Metavanadate solution**
   Dissolve 0.2 g of ammonium metavanadate (NH_4VO_3) in 25 ml hot distilled water. Cool and add 25 ml of 70% $HClO_4$.

6. **Molybdovanadate reagent**

   Mix 20 ml of molybdate solution, 25 ml of metavanadate solution and dilute to 100 ml with distilled water.

7. **Potassium orthophosphate monobasic (KH_2PO_4) standard**

   Stock (2 mg P/ml): Dissolve 0.8788 g KH_2PO_4 and dissolve to 100 ml in distilled water

   Working (0.1 mg P/ml): Dilute 0.5 ml of the stock into 100 ml of distilled water

Procedure

Weigh accurately 2±0.2 g of tissue sample and dry overnight in a hot air oven at 100°C. Place the sample in a muffle furnace to ash at 550°C for 4 h. Cool the content and add 20 ml of HCl (1+3), few drops of HNO_3 and boil. Cool the content, transfer to a 100 ml volumetric flask, dilute to a volume with distilled water, and filter. Transfer 1.0, and 2.0 ml of the filtrate in two test tubes. Transfer 0.5, 0.8, 1.0, 1.5, and 2.0 ml of the orthophosphate working standards to a series of test tubes. Transfer 2.0 ml of distilled water in a test tube to serve as a blank. To all the test tubes, add 2.0 ml of molybdovanadate reagent, and make up the volume to 10 ml with distilled water. Mix the content well and allow for 10 min. Measure the absorbance (O.D) at 420 mm in a spectrophotometer. Construct a calibration curve by plotting the absorbance value on the X-axis and orthophosphate (mg P) content on the Y-axis. Determine the phosphate content (mg P) in the sample solution from the graph.

Calculations

$$\text{Phosphorus, \%} = \frac{C}{\text{vol. (ml) of aliquot}} \times \frac{100 \text{ ml}}{\text{weight (g) of fish}} \times \frac{1}{10}$$

where,

C = phosphorus content (mg P) in sample; 10 = factor to change mg P/g to % p/g.

P_2O_5 , % = % P × 2.291 (2.291 = Conversion factor of P to P_2O_5)

Na_2HPO_4, % = % P × 4.583 (4.583 = Conversion factor of P to Na_2HPO_4)

Therefore, the mean and the standard deviation of the natural phosphorus content for a given species need to be established before the analysis by determining the "total added phosphate".

6.2.3. Titration methods

Calcium content can be determined by the titration methods in any biological material.

Calcium oxalate precipitates out by the addition of excess ammonium oxalate from an ammonium acetate solution buffered to a pH of 4.5-5.0. The precipitate, when dissolved in dilute H_2SO_4, liberates oxalic acid. Permanganate titration determines the oxalic acid, which gives the amount of calcium present.

Reagents

1. Saturated NH_4 oxalate solution, 6%
2. Dilute NH_4 OH (1:4)
3. Dilute H_2SO_4 (1:4)
4. Methyl red, 0.5 % in ethanol
5. $KMnO_4$ solution, 0.05 N

Ashing

Weigh accurately 10±1.0 g of the dried sample in a silica dish and ash at 500 – 550°C in a muffle furnace, until white. Cool the residue, add 5 ml of conc.HCl, and mix well. Evaporate the content to dryness in a water bath. Then, add 5 ml of conc.HCl, and boil for 2 min. Filter the content through a filter paper, wash with hot water, and make up the volume to 50 ml with distilled water. Use this solution for the estimation of minerals such as Ca, P, Fe, etc.

Titration

Transfer 10 ml of ash solution into a conical flask and dilute to 50 ml with distilled water. Measure 50 ml of distilled water in another conical flask as blank. Boil the contents in the flask and add 10 ml of saturated NH_4 oxalate solution and a drop of methyl red indicator. Cool the content and add dilute NH_4OH until the color of the solution is pink (pH 5.0). Leave the flask at room temperature for 4 h. Then, filter the solution through a Whatman No.4 filter paper, without disturbing the precipitate. Wash the flasks with distilled water and pour over the precipitate. Wash the precipitate until it is oxalate free. Transfer the funnel over another conical flask and break the tip of the filter paper with a clean rod. Wash the precipitate with hot water into the conical flask and add 10 ml of dil. H_2SO_4. Heat the flask to 90°C and add 50 ml of hot water and filter paper into the solution. Titrate the content against 0.05 N $KMnO_4$ solution. The appearance of a pale pink color that persists for 2 sec is the endpoint. Note down the volume of 0.05 N $KMnO_4$ consumed for the calculation of calcium content. Use this solution for the estimation of minerals such as Ca, P, Fe, etc.

Calculation

1 ml of 0.05 N $KMnO_4$= 1 mg of calcium

$$\text{Calcium (mg/100 g)} = \frac{\text{T.V} \times \text{N} \times 100 \times \text{D.F}}{\text{W} \times 0.05}$$

where, T.V – T.V of sample – T.V of blank, ml; N – Normality of $KMnO_4$; D.F – dilution factor; W – Wt. of the sample taken for ashing, g; 0.05 – calcium equivalence; 100 – percentage conversion

6.2.4. Redox reaction methods

Many analytical methods used are based on coupled reduction-oxidation (redox) reactions in mineral estimation. Analysts design a coupled reaction system such that one of the half-reactions leads to a measurable change in the system for use as an end-point, *e.g.*, a color change. One such reaction involves the mineral of analysis (*e.g.*, X = analyte), whereas the other an indicator (*e.g.*, Y = indicator). For example, permanganate ion (MnO_4^-) gives a deep purple color (oxidized form), while manganous ion (Mn^{2+}), a pale pink color (reduced form). Therefore, permanganate used as an indicator of many redox reactions. Calcium or iron content is determined by titration with a solution of potassium permanganate, the endpoint corresponding to the first change of the solution from pale pink to purple.

6.2.4.1. Iron estimation

Iron content determined based on the quantity of permanganate solution of known molarity that's required to succeed in the end-point. Potassium permanganate titrated with the aqueous solution of ash. When Fe^{2+} remains in the food, MnO_4^- converts Mn^{2+} and leads to the formation of a pale pink solution. Once all Fe^{2+} convert to Fe^{3+}, MnO_4^- remains in solution and leads to the formation of a purple color, the end-point.

Iron present on treatment with potassium thiocyanate in the presence of sulphuric acid gives a red color product having a maximum absorbance at 540nm.

Reagents

1. Potassium thiocyanate, 3N

   Dissolve 73 g of potassium thiocyanate in 250 ml of distilled water.

2. Saturated potassium persulphate solution

3. Conc.H_2SO_4

4. **Ferrous ammonium sulfate standard solution**

   Stock (1 mg/ml): Weigh accurately 70.2 mg of ferrous ammonium sulfate and dissolve in a little amount of water. Add 1 ml of conc.H_2SO_4, mix well and make up the solution to 100 ml with distilled water.

   Working (10 µg/ml):Pipette out 10 ml of the stock and make up to 100 ml with distilled water.

Procedure

Ash the sample first, before the estimation of calcium. Transfer 1, 2, 3, 4, and 5 ml of working standard ferrous ammonium sulfate solutions in a series of test tubes. Transfer 1 ml of distilled water, to serve as a blank. Transfer 0.5 ml and 1.0 ml of ash solution in another two test tubes. Make up the volume to 7.7 ml with distilled water in all the tubes. Add 0.4 ml of saturated potassium persulfate solution, 0.3 ml conc.H_2SO_4, and 1.6 ml of 3N potassium thiocyanate and mix. Draw a standard graph by plotting iron content on the X-axis and absorbance on the Y-axis. Calculate the iron content in the sample from the graph.

Calculation

$$\text{Iron (mg/100g)} = \frac{C \times 100 \times D.F}{W \times 1000}$$

where, C-Conc. of iron, µg; 100 – percentage conversion; D.F – dilution factor; W – wt. of sample taken for ashing, g; 1000 - µg to mg conversion

6.2.5. Precipitation titration methods

Precipitation titration is employed when any one product of a titration reaction is an insoluble precipitate. Mohr method for chloride analysis is the commonly used titrimetric method in the food industry. In this method, the sample in aqueous solution is analyzed by titration with silver nitrate containing chromate indicator. This reaction happens as the interaction between silver and chloride is much stronger than that between silver and chromate. Silver ion thus reacts with the chloride ion to form AgCl, until the entire chloride ion is exhausted. Any further addition of silver nitrate leads to the formation of silver chromate, which is an insoluble orange-colored solid precipitate. The end point of the reaction is the first appearance of an orange color. The volume of silver nitrate solution of known molarity required to reach the endpoint determines the concentration of chloride in solution.

6.2.6. Atomic spectroscopy methods

Determination of mineral contents by atomic spectroscopy is more sensitive, specific, and quicker than traditional methods.

6.2.7. Inductively coupled plasma analysis

Inductively coupled plasma mass spectrometry (ICP-MS) detects metals, several non-metals, and even isotopes at a low concentration. In this method, the plasma ionizes the sample into polyatomic ions for detection by mass analyzer. The method is rapid, sensitive, and precise than atomic mass spectroscopy.

Sample digestion

Weigh accurately 0.250 ± 0.05 g of sample in a digestion vessel and add 2.5 ml of 65% nitric acid (Suprapure). Place the vessel in a fume hood chamber for 1 h for pre-digestion of the sample. Then, add 7.5ml of MilliQ water to the pre-digested sample and perform the final digestion in a microwave by following a gradient temperature program. The program consists of ramping the temperature from ambient to 160°C at the rate of 20°C min^{-1} at 1000W, followed by a final hold at 160°C for 20 min. After the digestion, cool the mixture to below 90°C. Transfer the digested sample into the separate polypropylene centrifuge tube and make up the volume to 25ml using MilliQ water.

ICP-MS analysis

Perform the ICP-MS analysis to determine the concentration of metals. Use a multi-elemental standard mix as a certified reference compound. Calibrate the instrument first using the serially diluted multi-element standard from 0.1 to 25 ng/L. Use yttrium or other recognized metal as the internal standard. Set the radiofrequency (RF) generator and RF power condition at 27MHz and 1000W, respectively. Fix the plasma gas flow, nebulizer gas flow, and auxiliary gas flow rate at 15 L min^{-1}, 0.9 L min^{-1}, and 0.8 L min^{-1}, respectively. Determine the limit of detection (LOD) and limit of quantification (LOQ) values. LOD detects the lowest concentration of metal in a standardized condition, while LOQ quantifies the lowest concentration. Construct a calibration curve by plotting the concentrations of heavy metals (0.1, 1, 5, 10, 20, and 25ng/L) in the X-axis and the response factor on the Y-axis. Determine the concentration of each metal or mineral from the graph.

Accuracy of the method

Accuracy is the recovery percentage, defined as the confidence level between the reference value of the metal in the sample and the value obtained by the standard analytical method. The accuracy check with certified reference material (CRM) monitors the internal quality assurance check. The recovery percentage is expressed as mean±SD. As per the European Union requirements (EC/SANTE/ 11813/2017), the recoveries for all the elements should be within 80–120%, with standard deviation (SD) of<10%.

CHAPTER 7

SENSORY AND PHYSICAL QUALITY ANALYSIS

Freshness is the most important in the assessment of aquatic food quality. Quality of fish or shellfish degrades after the death as a result of autolytic and microbial degradation. Chemical reactions occur in aquatic food include changes in protein and lipid fractions, the formation of biogenic amines and hypoxanthine, and microbiological spoilage. As a result of these events, the sensory quality of aquatic food deteriorates. Aquatic food rich in PUFA's are susceptible to lipid oxidation and lead to the development of off-flavor, and off-odors called oxidative rancidity. Several methods are available for the evaluation of freshness and spoilage of aquatic food. Nevertheless, no single instrumental method is reliable for their assessment. They are sensory, non-sensory, and statistical methods. Non-sensory comprises of chemical or biochemical methods, physic-chemical, and microbiological method that can be applied by commercial aquatic food processing companies, and researchers to ensure their quality.

7.1. SENSORY QUALITY

The sensory evaluation provides rapid measurements of freshness. Sensory evaluation is defined as the scientific discipline to evoke, measure, analyze, and interpret reaction characteristics to food, perceived through the senses of sight, smell, taste, touch, and hearing. Sensory characteristics of whole fish are visible to consumers. Consequently, sensory evaluation continues to be the most satisfactory way of freshness assessment. There are two types of sensory methods, namely subjective and objective. The subjective assessment is the assessor's preference for a product estimated using adjectives such as like/dislike or good/bad. This method applies to market research and product development. The objective assessment is done based on organoleptic changes occurring in fish/shellfish on storage. The objective scoring scheme requires trained expert judges and panel. The assessors use their appropriate senses (sight, smell, taste, and touch) individually to determine each sensory characteristic, based on the defined grade standard, appropriate for the food. Sensory methods are also fast and nondestructive unless fish/shellfish is cooked, and is the most commonly used method in quality control.

7.1.1. EU Freshness Grading

EU freshness grading was introduced in the Council Regulation No. 103/76 (for fish) and 104/76 (for crustaceans), and updated by the Council Decision No. 2406/96 (for fish, some crustaceans, and only one cephalopod, the cuttlefish). EU scheme is commonly accepted in the EU countries for freshness grading of market fish within the Union and generally carried out by trained personnel in the auction site. Whole and gutted fish assessed in terms of acceptance of skins, eyes, gills, surface slime, belly cavity, odor, and texture of fish. There are four quality levels in the EU scheme, E (excellent), A (good quality), B (satisfactory quality), and C (unfit). Level E is the highest quality, and below level B, the fish is discarded or rejected for human consumption.

EU freshness grades

	E	A	B	C
Skin	Bright; shining; iridescent (not redfish) or opalescent; no bleaching	waxy; slight loss ofbloom; very slight bleaching	dull; some bleaching	dull; gritty; marked bleaching and shrinkage
Outer slime	transparent; watery white	milky	yellowish-grey;y some clotting	ellow-brown; very clotted and thick
Eyes	convex; black pupil; translucent cornea	plane; slightly opaque pupil; slightly opalescent	slightly concave; grey pupil; opaquecornea	completely sunken; greypupil, opaque discolored cornea
Gills	dark red or bright red; mucus translucent	red or pink; mucus slightly opaque	brown/grey and bleached; mucus opaque and thick	brown or bleached; mucus yellowish grey and clotted

[Table Contd.

Contd. Table]

		E	A	B	C
Peritoneum (ingutted fish)		glossy; brilliant; difficult to tear from flesh	slightly dull; difficult to tear from flesh	gritty; fairly easy to tear from flesh	gritty; easily torn from flesh
Gill and internal odors	all except plaice	fresh; sea weedy; shell fishy	no odor; neutral odor; trace musty, mousy, milky, caprylic, garlic orpeppery	definite musty, mousy, milky, caprylic, garlic orpeppery; bready; malty; beery; lactic; slightly sour	acetic; butyric; fruity; turnippy; amines; sulphide; faecal
	plaice	fresh oil; metallic; fresh cut grass; earthy; peppery	oily; sea weedy; aromatic; trace musty, mousy orcitric	oily; definite musty, mousy or citric; bready; malty; beery; slightly rancid; painty	muddy; grassy; fruity; acetic; butyric; rancid; amines; sulphide; faecal

This method has some disadvantages like the requirement of trained and experienced persons, lack of differentiation between species, and lack of information on the remaining shelf life of fish.

7.1.2. Quality Index Method

The quality index method (QIM) is an alternative to the EU scheme. The Tasmanian Food Research Unit in Australia originally developed the QIM (Bremner, 1985). This method is rapid and reliable to measure the freshness of whole fish held in ice, based on the significant sensory parameters like skin, slime, eyes, belly, odor, gills for raw fish (Branch and Vail, 1985). By assessing the characteristics listed on the sheet, an appropriate demerit point score is recorded (from 0 to 3). The sensory scores of all characteristics summed up to provide the overall score. A quality index (QI) close to 0 indicates very fresh fish, whereas a higher score depicts spoiled fish. There is a linear correlation between the sensory quality expressed as a demerit score and storage life on ice, which makes it possible to predict remaining storage life on ice. This method is relatively fast, non-destructive, and specific for species. The QIM is more suitable for the early stage of storage of fish where other instrumental methods are not accurate.

QIM scheme for Sensory Evaluation of Herring

	Quality Parameter	Description	Score
Whole fish	Appearance of skin	Very shiny	0
		Shiny	1
		Matte	2
	Blood on gill cover	None	0
		Very little (10-30%)	1
		Some (30-50%)	2
		Much (50-100%)	3
	Texture on loin	Hard	0
		Firm	1
		Yielding	2
		Soft	3
	Texture on belly	Firm	0
		Soft	1
		Burst	2
	Odour	Fresh seaweedy odour	0
		Neutral	1
		Slight off odour	2
		Strong off odour	3
Eyes	Appearance	Bright	0
		Somewhat lusterless	1
	Shape	Convex	0
		Flat	1
		Sunken	2
Gills	Color	Characteristic red	0
		Somewhat pale, matte, brown	1
	Odour	Fresh, seaweedy, metallic	0
		Neutral	1
		Some off odour	2
		Strong off odour	3
Total			20

Developed by Nielsen and Hyldig (2004) and modified from Jonsdottir (1992).

QIM Eurofish manual contains QIM schemes in 11 European languages for 13 fish species and information on their usage (Martinsdottir *et al.*, 2001). The QIM is rapid, easy to perform, non-destructive, and requires only short training.

It is applicable as a tool in production planning and quality warranty work. Rapid PC based QIM is available on the internet (http:www.dfu.min.dk/QIMRS/ qim_0202.htm). QIM schemes are available for deen water shrimp, farmed Atlantic salmon, fresh cod, whole fjord shrimp, herring, brill, haddock, plaice, pollock, redfish, (peeled) shrimp, sole, and turbot in EU.

7.1.3. Torry scheme

Torry Research Station developed the Torry scheme for usage by expert and trained judges to assess the fish freshness. This method is the most comprehensive scoring scheme for fish (Shewan *et al.*, 1953). Original or modified forms of this method are available. Torry Scheme is a descriptive 10 point scale, applicable to assess the freshness of lean, median fat, and high-fat fish species. The panelists evaluate the odor and flavor of cooked fish fillets also. The score ranges from 10 to 3. The freshness score 10 indicates a very fresh fish, while 3, the spoiled ones. Fish with an average score of 5.5 is fit for consumption.

Torry scheme for Freshness Evaluation of Cooked Cod Fillets

Odor	Flavor	Score
Initially weak odour of sweet, boiled milk, starch, followed by strengthening of these odours	Watery, metallic, starchy. Initially no sweetness by meaty flavours with slight sweetness may develop	10
Shellfish, seaweed, boiled meat	Sweet and meaty characteristic	9
Loss of odour, natural odour	Sweet and characteristic flavours but reduced in intensity	8
Wood shavings, wood sap, vanillin	Neutral	7
Condensed milk, boiled potato	Insipid	6
Milk jug odours, reminiscent of boiled clothes	Slight sourness, trace of "off" flavours	5
Lactic acid, sour milk, TMA	Slight bitterness, sour, "off flavours", TMA	4
Lower fatty acids (e.g. acetic acid or butyric acids) decomposed grass, soapy; turnip, tallow	Strong bitterness, rubber, slight sulfide	3

Developed by Shewan *et al.*, (1953)

7.1.4. Quantitative Descriptive Analysis

Quantitative descriptive analysis (QDA) is used by a trained panel to analyze the sensory attributes of products such as texture, color, and flavor. QDA provides a detailed description of all flavor characteristics, qualitatively and quantitatively.

A broad selection of reference samples made available to the trained panelists, and they use them for creating terminology that describes all aspects of the product. Descriptive words should be carefully selected, and the panelists trained should agree with the terms. Objective terms preferred over the subjective terms. The words for describing the odor and flavor of the fish/shellfish categorized into two groups, namely positive and negative sensory parameters based on whether fish/shellfish is fresh or spoiled.

Objective sensory methods are essential for quality control and estimation of shelf life of aquatic food. However, sensory methods are time-consuming, expensive, require trained personnel. Therefore, instrumental methods are needed to satisfy the quality testing of aquatic foods by the processing industry.

Quantitative Descriptive Analysis of farmed Nile tilapia (*Oreochromis niloticus*) eviscerated in ice

Appearance attributes	Definition
Colour of flesh	Colour going from white to light brown during storage period, not considering dark flesh
Brightness	Limpidity of colour, varying from opaque to bright, represented on the scale by "no brightness" to "a lot of brightness", accordingly
Orange pigment	Clear pigment, associated to fat oxidation
Odour attributes	**Definition**
Characteristic of fresh water fish	Strong fresh water fish odour; fresh water algae
Characteristic of sea water fish	Odour associated with fish stored for a long time in ice or beginning to deteriorate; sea smell
Rancid	Odour associated to deteriorated fat
Taste attributes	**Definition**
Characteristic of fresh water fish	Strong fresh water fish taste; fresh water algae
Characteristic of sea water fish	Taste associated with fish stored for a long time in ice or beginning to deteriorate
Bitter	Taste associated with rancidity – deteriorated fat (not consider dark flesh bitter taste)
Texture attributes	**Definition**
Softness	Force necessary to tear the flesh with the first bite
Juiciness	Amount of humidity in the mass liberated during mastication

Developed by Rodrigues *et al.* (2016)

7.2. PHYSICAL METHODS

7.2.1. Texture analysis

Texture analyses of food are necessary for research, quality control, and product development. The fish muscle becomes soft or mushy due to autolytic degradation or tough on frozen storage. Texture parameters include hardness, springiness, and chewiness of the food. Hardness is an important textural parameter, as it correlates well with the sensory assessments. Texture Profile Analysis is an imitative test that compresses the sample twice, mimicking the action of the jaw. The force-deformation curve obtained determines a number of texture parameters.

7.2.1.1. Measurement of Hardness by Texture Profile Analyzer

Use a flat-ended plunger or spherical probe for the analysis of whole fish, fillets, and mince. Place the probe or plunger first in contact with the sample. The software automatically records the thickness of the sample. Compress the sample to 80% of their original height at 2 mm/ sec speed using the cylindrical-shaped piston of 3 mm diameter for the first cycle. Once again, compress it within 5 sec after the first cycle. Record the force-deformation curve by the instrument. Express the hardness using the peak force of the first compression cycle in N. Repeat each measurement at least six times.

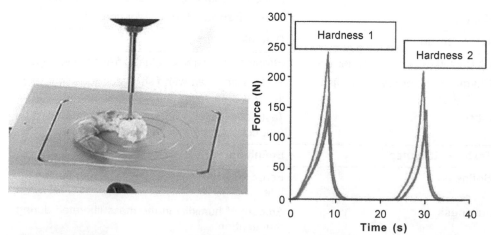

Picture for Hardness measurement in shrimp

Force- deformation curve due to hardness

7.2.2. Torry meter

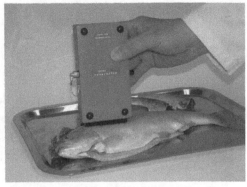

Torry Research Station in Aberdeen of Scotland developed the fish freshness meter "Torrymeter". Dielectric properties of fish used for the determination of fish freshness. Dielectric properties of fish skin and muscle alter systematically as tissue components degrade during spoilage. These changes occurring at a micro level are related to alterations in appearance,

Picture for Torrymeter testing freshness of fish

odor, texture, and flavor during spoilage and used a quality indicator. A linear relationship exists between torrymeter readings and sensory attributes for many fishes such as cod, Baltic herring, hake, blue whiting, flounder, mackerel, whole iced gilthead sea bream, and farmed Senegalese sole. Torry meter readings are significantly affected when fish is washed with seawater, as the ions interfere with the electrical properties of the skin. The presence of lipids also affects the dielectric properties of fish. Loss of skin, muscle integrity, and deterioration caused by bruising during harvest and packaging gave variable values.

7.2.3. Intellectron Fischtester VI

Intellectron Fischtester VI (Germany) measures the electric properties like resistance, conductivity, and capacitance of the fish flesh similar to Torrymeter. Dielectric properties can change after the death of the fish due to the disruption of the cell membranes by autolytic degradation. This method is based on conduction through skin and therefore works only on whole fish and fillets with skin. Mechanical abuse and freezing can affect the readings. Readings gives an objective measure on the state of freshness/spoilage together with sensory data.

7.2.4. RT-Freshtester

RT-Freshtester reflects the dielectric properties of fish like Torrymeter and Intellectron Fischtester VI. Reading decreases with storage time. It is fast and nondestructive and allows automatic grading of 60-70 fish/min. It needs calibration depending on sample preparation, season, fishing grounds, and fish handling procedures. It is unsuitable for frozen or thawed fish, partly frozen super chilled fish or fish chilled in RSW.

7.2.5. Cosmos

The Cosmos is an instrument developed by Japanese scientists to evaluate fish freshness. It is applied for the evaluation of fish quality by the determination of smell intensity. It is handheld, portable as well as rapid and nondestructive. It evaluates the freshness of fresh and chilled fish in the processing industry as well as in the fishing vessels. There is a strong correlation between the organoleptic and cosmos readings of six species of fish. Objective and quantitative evaluation of fresh and chilled fish quality is possible.

7.2.6. Electronic Nose

Odour is the main indication of fish/shellfish freshness. Element-Badvaki in Ireland developed an electronic nose called Fresh sense to assess the fish freshness. It is a rapid and nondestructive method to measure volatile compounds indicative of the spoilage odors in aquatic food. The electronic nose is a closed and static sampling system fitted with electrochemical gas sensors, which are sensitive to volatile compounds. The chemicals involved in fresh fish odors are long-chain alcohols and carbonyls, bromophenols, and N-cyclic compounds. Apart from this, short-chain alcohols and carbonyls, amines, sulfur compounds, and aromatic, N-cyclic and acid compounds produced by microbial activity and lipid oxidation during the storage of fish are also sensitive. The concentrations of these compounds are related to the degree of spoilage.

Different electronic noses are available for the measurement of fish freshness, namely, metal-oxide semiconductor gas sensors, electrochemical sensors (CO, H_2O, NO, SO_2 and NH_3), thickness-shear mode quartz resonators, semi-conductor dimethylamine (DMA) gas sensor, and prototype solid state-based gas sensor known as Fish Nose. There is a good correlation between the response of CO sensors and QIM methods. Data analysis gives the relationship between the sensor output patterns and the sample properties. The most frequently used methods are artificial neural works (ANNs), and chemometric analysis, such as the principle component analysis (PCA), and the partial least square regression (PLS-R).

7.2.7. Near Infrared Reflectance Spectroscopy

Near-infrared reflectance spectroscopy is rapid, speed, simplicity, and can measure numerous samples within a short time. It is non-destructive, easy to handle, and requires little training of operators. This method is applicable for quality assessment of frozen minced red hake and cod, caught by longline and gillnet, and also thawed chilled MAP cod fillets.

Fourier transform infrared spectroscopy (FTIR) is another technology that is a rapid, nondestructive, and suitable for on line industrial production chain. It requires too much handling of samples, causing changes in protein and muscle structure. Diffuse reflectance infrared Fourier transform (DRIFT) spectroscopy has advantages, as it is fast, simple, sensitive, inexpensive, and requires a small amount of sample. It was found useful in assessing the freshness and quality of sardine during iced storage.

7.2.8. pH

The pH is also an important parameter that shows depletion in fish tissue during storage. The postmortem pH varies from 5.5 to 7.1 depending on season, species, and other factors. Low initial pH is associated with higher stress before slaughtering, due to the depletion of energy reserves, mainly glycogen with the production of lactate. Since the activity of enzymes depends on pH, it affects reactions taking place during the storage of fish. A relatively low pH may cause a decrease in water-binding capacity of myofibrils, affect light scattering and appearance of fish, and promotes oxidation of myoglobin and lipids. The pH is determined by a pH meter, which has a glass electrode and calomel electrode. In most of the pH meters have combined a single electrode for measuring the pH. The pH of a solution defined as the negative logarithm of the hydrogen ion activity.

$$pH = - \log_a(H^+)$$

Reagents

1. Buffer solution 1: The pH of this solution is near 4.0.

2. Buffer solution 2: The pH of this solution is neutral 7.0

3. Buffer solution 3: the pH of this solution is near 9.2.

Procedure

Before the pH estimation, standardize the electrodes with suitable buffer solutions. Use pH 4.0 and pH 7.0 standard buffer solutions as standards for low pH foods. Use a minimum of two buffers to standardize the pH meter. Weigh 10 g of fish sample and homogenize with 50 ml of distilled water (pH 7.0). Dip the pH electrode in the sample solution and note down the readings.

CHAPTER 8

BIOCHEMICAL QUALITY ANALYSIS

Chemical and biochemical methods are reliable and accurate for the evaluation of seafood quality. They correlate well with sensory quality. The chemical compound determined should increase or decrease as microbial spoilage or autolysis progresses. The method used to evaluate fish freshness should combine with the measurements obtained from different methods and correlate with the findings of sensory analysis.

8.1. ATP and its breakdown products

Rigormortis occurs in postmortem muscle tissue and is associated with stiffness of muscle or flesh. This process results due to the breakdown of adenosine triphosphate (ATP). Nucleotide breakdown occurs due to autolytic enzymes and bacterial action. There is a correlation between nucleotide catabolism and loss of freshness. Initial stages of the reaction catalyzed by endogenous enzymes lead to the accumulation of adenosine diphosphate (ADP), and inosine monophosphate (IMP). Oxidation of hypoxanthine (Hx) to xanthine and uric acid is slower as the result of endogenous enzyme activity or microbial activity. The IMP is associated with fresh fish flavor, whereas inosine (Ino) and Hx reflect poor quality.

Adenosine triphosphate (ATP)
↓
Adenosine diphosphate (ADP)
↓
Adenosine monophosphate (AMP)
↓
Inosine monophosphate (IMP)
↓
Inosine (Ino)
↓
Hypoxanthine (Hx)
↓
Xanthine (Xa)
↓
Uric acid (Uric)

The concentration of ATP and its breakdown products serve as indicators of freshness in many fish species. K value proposed by Saito (1959) is the biochemical index for fish quality assessment based on nucleotide degradation. K value includes intermediate breakdown products, and it varies within species of fish.

$$K(\%) \frac{\ln o + Hx}{ATP + ADP + AMP + AMP + Ino + Hx} \times 100$$

Several methods are available for the analysis of single or a combination of nucleotide metabolites, but the HPLC method is the most reliable to calculate the K value.

8.1.1. K-value determination by HPLC method

Reagents

1. Perchloric acid, 0.6M
2. KOH, 1 M
3. Phosphate buffer solution (pH 6.8): Mix equal volumes of 0.04 M KH_2PO_4 and 0.06 M K_2HPO_4 in distilled water.

Extraction

Take 5 g of fish muscle tissue and homogenize with 25 ml of chilled 0.6 M perchloric acid at low temperature for 1 min. Centrifuge the homogenate at 6000 xg for 20 min at 4°C and filter. Take 10 ml of the supernatant, and neutralize to pH 6.8 with chilled 1M KOH immediately (visualized by the formation of KCl precipitate). Allow to stand at 0°C for 30 min, filter through a syringe filter (0.45m m), and store at -30°C for subsequent HPLC analysis.

HPLC analysis

Use reverse-phase C 18 column for the separation of nucleotides by isocratic elution using filtered phosphate buffer solution (pH 6.8). Maintain the flow rate at 1.5 ml/ min throughout the separation process. Inject 50ml of the sample into the HPLC. Identify the peaks of the sample by comparing it with the peak of the chromatogram of the standard mix. Quantify each nucleotide breakdown products by comparing the peak area of the samples with the peak area of the standards (Ryder, 1985).

8.1.2. K-value determination by UPLC method

Reagents

1. **Standard ATP degradation products :** Adenosine triphosphate (ATP), Adenosine diphosphate (ADP), Adenosine monophosphate (AMP), Inosine monophosphate (IMP), Inosine (I), Adenosine (A) and Hypoxanthine (Hx) from Sigma

2. Perchloric acid, 0.6M

3. KOH, 1M

4. **Phosphate buffer, pH 7.0:** Dissolve 0.04 M potassium dihydrogen phosphate and 0.06M of dipotassium hydrogen phosphate in 1 L of distilled water.

5. Acetonitrile (ACN)

6. MilliQ water

Preparation of standards

1. **Stock standard (10μM):** Weigh 27.56 mg of adenosine tri phosphate (ATP), 31.27 mg of adenosine di phosphate (ADP), 17.361 mg of adenosine mono phosphate (AMP), 19.61 mg of inosine mono phosphate (IMP), 13.41 mg of inosine (I), 13.35 mg of adenosine (A) and 6.805 mg of hypoxanthine (Hx) individually and make up to 5 ml with milliQ water.

2. **Working standard I (250μM):** Take 25μl from each stock and dilute to 1 ml with distilled water.

3. **Working standard II (100μM):** Take 10μl from each stock and dilute to 1 ml with distilled water.

Procedure

Take 5±0.5g of fish muscle tissue and homogenize with 25ml of chilled 0.6 perchloric acid at low temperature for 1 min. Centrifuge the homogenate at 6000xg for 20min at 4°C and filter. Take 10ml of the supernatant, neutralize to pH 6.8 with chilled 1M KOH immediately, which can be visualized by the formation of KCl precipitate. Allow the solution to stand at 0°C for 30 min, filter through a syringe filter (0.2μm) and store at -20°C for subsequent UPLC analysis.

UPLC analysis

Acquity UPLC system attached with UPLC B EHC 18 2.1x100mm column having a pore size of 1.7Mm used for the separation. Set the injection volume to

2 µl and the total run time to 15 min. The mobile phase consisted of phosphate buffer and acetonitrile : milliQ water (80:20). Set the flow rate to 0.2 ml/min. The gradient programming is given below:

	Solvent A Phosphate Buffer	Solvent B ACN: water (80:20)
Initial	100	Initial
9.50	100	0
11	50	50
15	50	50

Interpretation of results

Identify the amines by comparing the retention time of sample with that of the authentic working standards. Quantify the concentration of sample by comparing the peak area of sample with that of the standards.

8.1.3. Hypoxanthine estimation by spectrophotometric method

Hypoxanthine (Hx) formed due to the autolytic degradation of ATP is a useful indicator of fish freshness. Hx content increases with the time of storage to reach a level of 5 m moles/ g wet weight and subsequently declines or remains at that level. The content of hypoxanthine correlates well with sensory assessment, the flavor in particular. The limit of acceptability differ for each fish species, i.e., cod – 2 to 3 mm/g; herring – 2 to 2.5 mm/g; mackerel – 1 to 1.5 mm/g; shrimps - 2mm/g; squids- 2 to 4mm/g.

A protein-free extract of fish muscle made using perchloric acid is neutralized, and the perchlorate removed from it, as insoluble potassium salt. The hypoxanthine in the neutral solution converted by the enzyme xanthine oxidase to uric acid, which has an absorbance maximum at 290 nm.

Reagents

1. Perchloric acid, 0.6M
2. NaOH, 1M
3. KOH, 1M

4. **Potassium hydroxide – phosphate buffer**

   Dissolve 27.2 g of KH_2PO_4 in about 250 ml water and add 170 ml of 1.0M NaOH. Check that the pH is 7.6 ± 0.5 and adjust, if necessary. Add 557 ml of 1M KOH to the buffer solution and make up to 1 L with distilled water.

5. **Phosphate Buffer, pH (7.6),0.25M**

   Dissolve 17.0 g of KH_2PO_4 in about 250ml distilled water. Add 107 ml of 1.0M NaOH and check the pH. Adjust the pH 7.6±0.5, if necessary. Make the volume up to 500 ml, with distilled water.

6. **Phosphate buffer, pH (7.6) 0.05M**

   Dilute 0.25M phosphate buffer five folds with distilled water.

7. **Xanthine oxidase working solution**

   Commercially available stock enzyme solution has an activity of 3-4 EU/ml. Dilute this stock solution 50-folds with chilled 0.05 M phosphate buffer (Prepare fresh each day)

8. **Hypoxanthine standard solutions**

   *Stock solution (100mg/ml):*Dissolve 10 mg of hypoxanthine in about 70 ml of boiling water. Cool and make up to 100ml with distilled water. Make up fresh each week and store in a refrigerator.

   *Working solution:*Dilute the stock with distilled water to give working standards containing 5, 10, 15, 20, 25mg hypoxanthine in 1.0 ml. Check the concentration by measuring the optical density of the 10mg/1.0 ml solution at 250 nm such that it should be 0.786. (The molar extinction coefficient of hypoxanthine is 10.7×10^3 and molecular weight is 136.1).

Procedure

Weigh accurately 5.00±0.05 g of the tissue sample and homogenize with 50 ml of chilled 0.6M perchloric acid. Centrifuge at 5000 xg for 15 min and filter it through Whatman No. 1 filter paper. Take 5.0 ml of the filtrate and add 5.0 ml of KOH – phosphate buffer, chill, and filter again.

Solution A (Reaction mixture): Pipette out 0.5, 1.0, and 2.0 ml of filtrates into the test tubes and makeup to 2.0 ml with distilled water. Add 2.0 ml of 0.25 M phosphate buffer, and 0.5 ml of enzyme solution.

Solution B (Extract blank): Pipette out 0.5, 1.0, and 2.0 ml of filtrates into the test tubes and makeup to 2.0 ml with distilled water. Add 2.0 ml of 0.25 M phosphate buffer and 0.5 ml of distilled water.

Solution C (Enzyme blank): Prepare a blank containing 2.0 ml of 0.25 M phosphate buffer, 2.0 ml of distilled water, and 0.5 ml of enzyme solution.

Solution D (Buffer blank): Prepare a blank containing 2.0 ml of 0.25 M phosphate buffer and 2.5 ml of distilled water. Incubate all the mixtures at 35°C for 30 min. Measure the absorbance at 290 nm in the spectrophotometer. Calculate the NIL absorbance due to uric acid after correcting for the blanks.

$$AB = (A\text{-}D) - (B\text{-}D) \text{ - } (C\text{-}D)$$
$$= (A\text{-}B) - (C\text{-}D)$$

Calculation

The molar extinction coefficient of uric acid at 290 nm is 12.4×10^3. One mole of uric acid is equivalent to one mole of hypoxanthine in 4.5 ml of the reaction mixture:

$$\frac{AB \times 136.1 \times 10^6 \times 4.5}{12.4 \times 10^3 \times 10^3} = AB \text{ x } 49.4 \text{ mg}$$

Extract the hypoxanthine from fish muscle tissue with 54 ml of 0.6M perchloric acid, assuming that fish contains 80% water. This extract is diluted 2 folds in step prior to analysis.

Concentration of hypoxanthine in a 5g sample of fish is

$$\frac{AB \times 49.4 \times 54 \times 2 \times 100}{V \times 1000 \times 5} = \frac{AB \times 119}{V \text{ mg} / 100g}$$

where, V is the volume of extract taken to prepare the reaction.

8.2. BIOGENIC AMINES

The biogenic amines estimation is a reliable method of measuring the quality of fish, depending on the species. The formation of biogenic amines results from microbial degradation during the later storage of fish, and their concentration increases with storage time. The microbial decarboxylation of specific free amino acids in fish or shellfish tissue forms the biogenic amines. The significant biogenic amines produced postmortem in fish and shellfish are histamine, putrescine, cadaverine, tyramine, tryptamine, 2-phenylethylamine, spermine, spermidine, and agmatine. Spoilage bacteria produce biogenic amines toward the end of shelf life of a fish, and hence their levels are considered as indices of spoilage rather than freshness.

Histamine is the potentially hazardous and causative agent of histamine intoxication, among the biogenic amines. Other biogenic amines are putrescine and cadaverine that enhance the toxicity of histamine. The legal limits for histamine are set 5 mg/100g and 20 mg/100g fish by USFDA and EU, respectively. Biogenic amine content in fish depends on the species, free amino acids, presence of decarboxylase positive microorganisms, capture, and stomach contents at death.

Tyrosine produces tyramine, histidine yields histamine, arginine becomes putrescine, lysine produces cadaverine, tryptophan forms tryptamine, and phenylalanine yields 2-phenylethylamine, through decarboxylation reactions. Putrescine, an intermediate in the metabolic pathway, leads to the formation of spermidine and spermine. A quality index (QI) and biogenic amine index (BAI) methods are the two methods used for the determination of biogenic amines, as proposed by Mietz and Karmas (1977) and Veciana-Nogues *et al.* (1995), respectively

$$QI = \frac{histamine + putrescine + cadaverine}{1 + spermidine + spermine}$$

BAI = histamine + putrescine + cadaverine + tyramine

The QI is determined based on the increases in putrescine, cadaverine, and tyramine and decreases in spermine and spermidine, whereas, the BAI is on the increases in histamine, putrescine, cadaverine, and tyramine, during the storage of fish.

There are various analytical techniques to determine the concentration of biogenic amines, such as thin-layer chromatography (TLC), HPLC, GC, ELISA, capillary zone electrophoresis, and biosensors. Among these techniques, HPLC is more sensitive, reliable, and reproducible.

8.2.1. TLC method

Amines extracted in trichloroacetic acid get converted by dansyl chloride into fluorescent dansyl derivatives, which are visualized under UV and quantified using a densitometer at 356 nm.

Reagents

1. TCA, 5%

2. NaOH, 4 N

3. Phosphate buffer (pH 9, undiluted)

4. **Dansyl reagent**

   Dissolve 50 mg of dansyl chloride (5-dimethyl amino naphthalene 1-sulphonyl chloride) in 10 ml of acetone (Prepare fresh before use)

5. **Biogenic amine standard**

   Stock (2mg / ml): Dissolve 0.20 mg of each amine (histamine dihydro chloride, cadaverine dihydrochloride, putrescine dihydrochloride and tyramine dihydrochloride) in 100 ml of 5% TCA

   Working (0.2mg / ml):Pipette out 0.5 ml of the stock and make up the volume to 50 ml with 5% TCA

Procedure

Dissolve 10 g silica gel G (with gypsum binder) in 30 ml of distilled water to prepare silica slurry for making one 20 x 20 cm plate. Pour the slurry over the glass plate and spread with the glass-rod uniformly to prepare the plate. Allow the plate to air-dry and place it at 110°C for 1 h for activation before use. Weigh 10 g of tissue sample, add 30 ml of hot 5% (80-90°C) TCA and homogenize for 2 min. Centrifuge at 3000 rpm for 10 min and filter through the Whatman No. 1 filter paper. Transfer 1 ml of the extract to a stoppered test tube, to which add a drop of 4 N NaOH, 1 ml of phosphate buffer, and 2 ml of dansyl reagent. Close the tube, mix, wrap with aluminum foil, and place in a thermostatic oven maintained at 50°C for 1 h. Similarly, derivatize 1 ml of each amine working solution to respective dansyl derivatives. Mix equal quantities of each dansyl standard to obtain standard amine mixture.

Place 25 ml of dansyl amines as spots on the activated silica gel plate. Simultaneously, place 25 ml of standard amine mix as a spot alongside the sample in the silica gel plate. Develop the chromatogram using chloroform: triethylamine (100: 20) to separate the amines. After development, spray the plate with triethanolamine: isopropanol (8: 2) to enhance the fluorescence. View the plate under UV light at 254 nm, calculate the Rf values, and compare with the intensity of fluorescent spots of the authentic standards. Scan the TLC plates in a TLC scanning densitometer at 356 nm. Calculate the concentration of individual amines with the help of area intensities of the authentic standards (Jeyshakila *et al.*, 1999).

8.2.2. HPLC method

Separation of biogenic amines using dansyl amine derivatives was described by Rosier and Petegham (1988) by gradient elution program.

Apparatus

- HPLC system equipped with an automated gradient controller and pumps
- Column : C-18µ Bondpak RP, 300 x 3.9mm,
- Detector : UV set at 254 nm
- Mobile solvents: Methanol and water

Procedure

Use the dansylated amines samples and standards prepared for the TLC analysis of biogenic amines. Equilibrate the column with the methanol/ water mixture (70:30). Adjust the instrument conditions to give a full-scale response by keeping AUFS at 0.1 with a noise rejection at 50 and area rejection at 100. Inject 5-10 µl aliquot of the amine standard mixture after filtration onto the column. Run the chromatography as per the following gradient elution program.

Time (Min)	Solvent A (%) Water	Solvent B (%) Methanol	Flow rate (ml/min)
Initial	30	70	1.00
3.5	25	75	1.00
7.0	20	80	1.00
12.0	0	100	1.00

Generate the chromatographic data processed by the data module integrator. Then, inject 5-10 µl aliquot of the amine sample after filtration onto the column. Calculate the amount of biogenic amines in the sample by comparing the peak area intensities with those from the standard mixture.

8.2.3. Biogenic amine analysis by UPLC method

Reagents

1. Standard biogenic amines: Histamine HCl, Cadaverine HCl, Putrescine HCl, and Tyramine HCl from Sigma
2. HCl, 0.1 M

3. Perchloric acid, 0.6M
4. NaOH, 0.6M
5. NaOH, 2M
6. Dansyl chloride, 1%
7. Saturated sodium bicarbonate
8. Ammonia
9. Acetonitrile
10. MilliQ water

Preparation of standards

1. **Stock standard (50µM):** Weigh 133.62 mg of histamine HCl, 87.55mg of cadaverine HCl, 80.53 mg of putrescine HCl, and 83.62 mg of tyramine HCl individually and make up to 10ml with 0.1 M HCl in 10ml volumetric flask
2. **Working standard (0.2µM):** Take 20µl from each stock and dilute to 1 ml with distilled water.

Extraction of biogenic amines

Weigh 5±0.5g of fish sample and homogenize with 12 ml of 0.6M perchloric acid. Repeat the extraction thrice, filter and centrifuge at 4000xg for 10 min at 4°C. Transfer 1.5ml of the supernatant to an Eppendorf tube and neutralize with 0.6M of NaOH. Centrifuge the neutralized sample once again at 9.000xg for 10 min at 4°C.

Dansylation of biogenic amines

Take 1 ml of the neutralized extract, add 0.2 ml of 2M NaOH, 0.3ml of saturated NaHCO$_3$ and 2ml of 1% dansyl chloride and place in a water bath at 55°C for 45 min. Then, add 0.1ml of ammonia solution and keep in dark condition for 10 min. Next, add acetonitrile to makeup the final volume to 5ml and centrifuge at 9000xg for 5 min at 4°C. Filter the supernatant through PVDF syringe filter (0.2µm) beofer the injection into UPLC.

UPLC analysis

Acquity UPLC system attached with UPLC B EHC 18 2.1x100mm column having a pore size of 1.7Mm used for separation of the biogenic amines. Set the injection volume to 2 µl and the total run time to 13 min. The mobile phase consisted of milliQ water and acetonitrile. Set the flow rate to 0.4 ml/min. The gradient programming is given below:

	Solvent A MilliQ water	Solvent B Acetonitrile
Initial	50	50
5	50	50
8	15	85
12	50	50

Interpretation of the results

Identify the amines by comparing the retention time of sample with that of the authentic working standards. Quantify the concentration of sample by comparing the peak area of sample with that of the standards using the Empower software.

8.3. HISTAMINE

Decarboxylation of histidine amino acid by bacteria leads to the formation of histamine in fish. Histamine is also an indicator of decomposition and causes food poisoning termed as 'scombroid poisoning'. Fish producing histamine mainly include pelagic fishes like tuna, mackerels, carangids, etc. Later, even non-scombroid fishes such as sardines, herring, and mahi-mahi cause histamine poisoning. The natural level of histamine in fresh fish is less than 5 mg/100 g. Histamine can be determined in fish by fluorometric and chromatographic methods. Histamine can be eluted from fish by butanolic extraction as well as by ion exchange resin column before fluorimetric analysis.

8.3.1. Butanolic extraction method

Histamine extracted using perchloric acid is again partitioned n-butanol under alkaline condition. Extract containing histamine collected in acid-aqueous phase condensed with o-pthalaldehyde under alkaline condition emits fluorescence, which has an excitation wavelength maximum at 340 nm and an emission wavelength maximum at 425 nm.

Reagents

1. Perchloric acid, $HClO_4$, 4N
2. HCl, 0.1N
3. NaOH, 1N

4. NaOH, 0.1N

5. Sodium chloride

6. n-Butanol

7. Petroleum ether

8. ortho-opthaldialdehyde solution (OPT), 1 % in methanol (Prepare fresh daily)

9. Citric acid, 0.2M

10. Histamine standard solution

Stock (1000 ppm): Dissolve 16.55 mg of histamine dichloride in 10 ml of 0.1N HCl solution

Working(10 ppm): Pipette out 1.0 ml of the stock and make up the volume to 100 ml with 0.1N HCl solution in volumetric flask

Extraction

Weigh 5 g of fish tissue and homogenize with 25 ml of 0.4 N $HClO_4$ for 1 min. Centrifuge the homogenate at 5000 xg for 10 min, filter through Whatman No. 1 filter paper and makeup to 50 ml with 0.4 N $HClO_4$. Transfer 5 ml of the extract into 250 ml separating funnel, add 10 ml of distilled water and make alkaline with 5 ml of 1N NaOH. Then, add 1.5 – 2.0 g of NaCl (to saturation) and extract two times with 10 ml of n-butanol. Collect all the butanolic phases. Then, wash the butanolic phases with 10 ml of 1N NaOH and 2.0 g NaCl. Remove the aqueous phase and add 25 ml of petroleum ether and shake well. Discard the aqueous phase. Then, extract again the butanolic phase two times with 10 ml of 0.1N HCl. Collect the aqueous-acid phase (0.1N HCl) in a 50 ml volumetric flask and makeup to 50 ml with 0.1N HCl. Take 0.2, 0.4, 0.6, 0.8, and 1.0 ml of the histamine working standard in a series of test tubes and make up the volume to 5ml with 0.1N HCl. Take 5.0 ml of 0.1N HCl as blank in a separate test tube.

Condensation reaction

Transfer 2.0 ml of each standard, sample, and blank in separate test tubes. Add 4.0 ml of 0.1N NaOH and shake well. Then add 0.2 ml of OPT solution, shake to mix, and let them to stand for 5 min. Then, add 2.0 ml of 0.2 M citric acid to the tubes and shake well. Measure the fluorescence at the excitation wavelength of 340 nm and an emission wavelength of 425 nm in a spectrofluorometer. Draw the standard graph by plotting the concentration of the histamine (mg) on the X-axis and the fluorescent intensity (mV) on the Y-axis. Compute the concentration of histamine from the standard graph.

Calculation

$$\text{Histamine (ppm)} = \frac{C}{W} \times D.F$$

where, C – conc. of histamine, ppm; W – weight of the fish, g; D.F – dilution factor

8.4. TOTAL VOLATILE BASE NITROGEN

In fish and shellfish, total volatile base nitrogen (TVB-N) primarily includes trimethylamine (TMA), produced by spoilage bacteria, ammonia produced by the deamination of amino acids and nucleotide catabolites, and dimethylamine (DMA) produced by autolytic enzymes during frozen storage. Analyses of these compounds reflect later stages of spoilage rather than freshness. However, the European Commission on Fish Hygiene (Council Regulation No. 95/149/EEC of March 1995) specifies that TVB-N be used as a chemical check if the organoleptic examination indicates any doubt as to the freshness of the fish.

The level of TVB-N in freshly caught fish is generally between 5 - 20 mgN/ 100g. However, the levels of 30-35 mg N/100 g muscle are considered the limit of acceptability for ice stored cold-water fish. There are three different methods for estimation of TVB-N: EC reference method, Conway micro diffusion method, and steam distillation method. EC reference method involves preliminary deproteinization with perchloric acid. Distillation method involves direct distillation of fish with the addition of magnesium oxide. Conway method uses trichloric acid instead of perchloric acid for deproteinization. A good correlation exists between the three methods however, the direct distillation method is a rapid method for routine analysis.

8.4.1. Distillation method

Volatile bases liberated during distillation are absorbed by the acids, which is then quantified by back titration of excess acid against a standard acid.

Reagents

1. Perchloric acid, 6%
2. Boric acid, 2%

3. **Mixed indicator**

   Mix equal volumes of 0.066% of alcoholic methyl red and alcoholic bromocresol green solutions

4. HCl, 0.05N

Procedure

Weigh 10g of fish muscle tissue and homogenize with 50 ml of perchloric acid to precipitate the protein. Centrifuge the homogenate at 4000 xg for 15 min at 5°C and filter. Take 5 ml of supernatant and distill in Kjeldahl apparatus (as for protein analysis). Collect the distillate in 10 ml of boric acid containing 2-3 drops of mixed indicator in a conical flask. Titrate the contents against 0.05 N HCl until pink color appears. Simultaneously, distill 5 ml of distilled water in the Kjeldahl apparatus as a blank and proceed as above. The difference in the titer values of the sample and blank gives the actual titer value of the sample.

Calculation

$$\text{TVBN (mg \%)} = \frac{\text{T.V} \times 14 \times \text{N} \times 100 \times \text{D.F}}{\text{W}}$$

where, T.V- titer value of the sample, ml: 14 – mg of N_2 in 1 N HCl, N – Normality of HCl (0.05); 100 – percentage conversion, W – weight of the fish, g, D.F – dilution factor

8.5. TRIMETHYLAMINE

Trimethylamine oxide (TMAO) is broken down by the activity of TMAO demethylase into dimethylamine (DMA) and formaldehyde (FA). TMAO appears to be part of the fish and used in osmoregulation. TMA formed by the decomposition of TMAO due to microbial spoilage and enzymatic activity after the death of fish. TMAO reductase is a microbial enzyme that can reduce TMAO into trimethylamine (TMA). Seawater fish contain 1-100 mg TMAO/100 g tissue, whereas freshwater fish have 5-20 mg/100g. Marine teleosts contain 15 – 250 mg TMAO/100 g tissue, while marine elasmobranches have 1000 – 1800 mg TMAO/ 100 g. Fresh fish has a low amount of TMA of less than 1.5mg TMA/100 g, but the values increase during spoilage. Fish is stale when the TMA production is higher than 30 mg/100 g. Hence, TMA is used as a microbial spoilage indicator and not as an index of freshness. The permissible limit for TMA is 10-15 mg/100g fish.

There are many analytical methods for the measurements of TMA, DMA, or TVB-N: steam distillation, Conway micro diffusion titration, colorimetric, photometry, HPLC, GC, injection gas diffusion, and biosensors. Biosensor using flavin-containing monooxygenase type-3, and solid-state sensors using bromocresol green are few.

8.5.1. Conway micro-diffusion method

Volatile bases liberate when potassium carbonate (K_2CO_3) reacts with fish tissue. The liberated NH_3 is absorbed in acids and determined by titration of excess acids against a standard alkali.

Reagents

1. Sulphuric acid, N/70
2. Sodium hydroxide (NaOH), N/70
3. Trichloroacetic acid (TCA), 20%
4. TCA, 2%
5. Saturated potassium carbonate
6. **Neutral formaldehyde solution or 20% formaldehyde**

   Shake well 100 ml of commercial formalin (i.e. 40%) with 10 g of $MgCO_3$ until it becomes colorless and filter it. Then, dilute with 100 ml of distilled water to make it up to 20% formaldehyde.
7. **Tashiro's Indicator**

   Add 20 ml of 1% alcoholic solution of methyl red to 5 ml of 1% alcoholic methylene blue solution. Mix one volume of this stock solution, one volume of alcohol and 2 volumes of distilled water to get the indicator.

Procedure

Weigh 10 g of fish sample, add 20 ml of 20% TCA and homogenize. Filter using a Whatman No. 1 filter paper in a 50 ml volumetric flask. Repeat the TCA extraction, filter, and make up the volume of extract to 50 ml. Pipette out 1 ml of N/70 H_2SO_4 into the inner chamber of the Conway unit and add a drop of Tashiro's indicator. In the outer chamber, add 1 ml of TCA extract, 0.5 ml of neutralized formalin, and cover the unit with the greased cover glass, leaving little gap for adding potassium carbonate. Then, add 1 ml of K_2CO_3 in the outer chamber by rotation, cover the unit, and keep it overnight at room temperature or 37°C for 90 min. Then, remove the lid and titrate the contents of the inner chamber against N/70 NaOH using a micro burette, until the change of color. Conduct a blank using 1ml of 2% TCA solution.

Calculation

$$\text{TMA (mg/100g)} = \frac{V \times 0.2 \times 50 \times 100}{W}$$

where, V - difference in the titer values of blank and sample, ml; 0.2- nitrogen conversion factor; 50 - dilution factor; 100 - percentage conversion; W- weight of the fish, g

8.6. DIMETHYLAMINE

The TMAO is converted to DMA and formaldehyde by the enzyme TMAO demethylase during chilled or frozen storage of fish and when bacterial growth is inhibited. The formation of DMA and FA may cause severe quality changes or spoilage during prolonged frozen storage. The amount of DMA produced depends on species, storage temperature, and time. DMA is a spoilage index during the frozen storage of fish species.

8.7. Formaldehyde

The presence of formaldehyde in aquatic food is generally non-toxic. It can react with many chemical compounds such as amino acid residues, terminal amino groups, and low molecular weight compounds. It can denature the proteins, form cross-linkage with proteins, and reduce the solubility of myofibrillar proteins. Formaldehyde content in frozen seafood is a spoilage index, especially in lean fish. Food Safety and Standards Authority of India (FSSAI) fixes maximum permissible limits for formaldehyde in fish, including crustaceans, molluscan and echinoderms as 1 mg/kg for freshwater origin and 100 mg/kg for brackishwater or marine origin (FSSAI, 2019).

8.7.1. Spectrophotometric method

Protein free TCA extract neutralized using alkali reacts with Nash reagent consisting of ammonium acetate and acetylacetone to form a colored complex having an absorption maximum at 415 nm.

Reagents

1. TCA, 10% or perchloric acid, 6%
2. Sodium hydroxide solution, 45% and 1.0 M
3. **Double strength Nash reagent (DSNR)**

   Dissolve 150 mg of ammonium acetate and 2 ml of redistilled acetyl acetone in 500ml distilled water (This solution is stable for 6 months at 0°C).
4. **Formaldehyde standard solution, 38%**

   Stock: Dilute 5 ml of formaldehyde in 1000 ml with distilled water

   Working (3.8 ppm/ml): Dilute 5 ml of the stock solution to 500 ml with distilled water

Procedure

Weigh 10 g of fish tissue and homogenize with 30 ml of 10% TCA for 1 min. Centrifuge at 5000 xg for 15 min at 4°C and filter. Pipette out 5 ml of the supernatant to 15 ml of distilled water taken in a beaker. Titrate to pH 6 using 45% NaOH first, and then with 1.0 M NaOH solution, and makeup to 25 ml with distilled water, mix and filter. Pipette out 5 ml of the neutralized filtrate in a test tube. Pipette out 0.5, 1.0, 1.5, 2.0, 2.5, 3.0, and 3.5 ml of the working standard in a series of test tubes and makeup to 5 ml with distilled water. Pipette out 5 ml of distilled water in another test tube to serve as a blank. Add 5 ml of DSNR solution to all the test tubes, shake vigorously, and heat in a water bath set at 60°C for 5 min. Cool the test tubes and measure the absorbance (O.D) at 415nm in a spectrophotometer.

Calculation

$$\text{Formaldehyde (in ppm)} = \frac{C \times D.F}{W}$$

where, C - concentration of formaldehyde, ppm; D.F – dilution factor; W - weight of the fish, g

8.8. INDOLE

Indole is a decomposition product formed by the breakdown of tryptophan, amino acid present in shrimps. Indole is a 2, 3 benzopyrolle with a fecal odor, highly volatile and soluble in hot water, alcohol, ether, and benzene. Indole is a useful spoilage indicator in shrimps and crabs, and its content should not exceed 25mg

/100g. There are two methods for the estimation of indole: extraction and distillation of which, the latter gives more recovery.

8.8.1. Extraction method

Indole extracted using trichloroacetic acid reacts with Eldrich's reagent (p-dimethyl amino benzaldehyde) to form a red color compound having an absorption maximum at 570 nm.

Reagents

1. TCA, 10%

2. Petroleum ether (60-80°C)

3. **Eldrich's reagent**

   Dissolve 1.8 g of p-dimethyl amino benzaldehyde (p-DMAB) in 9 ml of conc. HCl in 50 ml volumetric flask and dilute with ethanol. (Prepare fresh daily)

4. **Indole standard**

   Stock (1mg /ml): Dissolve 100 mg of indole in 100 ml of petroleum ether

   Working (5mg/ml): Pipette out 0.5 ml of the stock and make up the volume to 100 ml with petroleum ether in volumetric flask

Procedure

Weigh 10 g of the tissue sample and homogenize with 20 ml of 10% TCA solution for 1 min. Centrifuge the homogenate for 10 min at 15,000 xg. Filter the supernatant through Whatman No. 1 filter paper under slight suction. Transfer the filtrate into 250 ml separating funnel and add 8 ml of petroleum ether. Shake well and allow the two layers to separate. Transfer the petroleum ether layer to another separating funnel and extract once again with 8 ml of petroleum ether. Combine the extract in the second separating funnel. Then, extract with 5 ml of Eldrich's reagent by vigorously shaking for 1 min. Take 1, 2, 3, and 4 ml of the indole working standard along with blank (petroleum ether) in separate test tubes and add 5 ml of Eldrich's reagent. Read the color at 570 nm. Draw a standard graph by plotting the concentration of the indole (mg) on the X-axis and the optical density (OD) on the Y-axis. Compute the concentration of indole in the sample from the standard graph.

Calculation

$$\text{Indole (mg / 100g)} = \frac{C \times D.F \times 100}{W}$$

Where, C - concentration of indole, mg; D.F - dilution factor; W - weight of sample, g; 100 - % conversion

8.8.2. Distillation method

Sample is steam distilled to release indole, then extracted using chloroform, and allowed to react with Eldrich's reagent (p-dimethyl amino benzaldehyde) to form a red-colored compound having an absorption maximum at 570 nm.

Reagents

1. Alcohol
2. 5% Hydrochloric acid: Dissolve 5 ml of HCl in 100 ml of distilled water
3. Saturated sodium sulfate
4. **Eldrich's reagent**

   Dissolve 0.4 g of p-dimethyl amino benzaldehyde (p- DMAB) in 5 ml of acetic acid, mix with 92 ml of phosphoric acid and make up to 100 ml. (Prepare fresh daily)
5. **Indole standard**

   Stock (1 mg/ml): Dissolve 100 mg of indole in 100 ml of alcohol

   Working (10 µg/ml): Pipette out 1.0 ml of the stock and make up the volume to 100 ml with alcohol in volumetric flask

Procedure

Weigh 5 g of the tissue sample and homogenize with 80 ml of alcohol. Transfer the content into a distillation flask, distill, and collect 50 ml of distillate. Transfer the distillate into a separating flask, add 5 ml of 5% HCl and 5 ml of saturated sodium sulfate. Extract with 25 ml of chloroform and shake well. Allow the layers to separate and collect the lower chloroform layer. Add 10 ml of the Eldrich's reagent to the chloroform extract, shake well, and allow the acid layer to separate. Transfer the acid layer into a 50 volumetric flask and dilute to mark with acetic acid. Measure the color at 590 nm in the spectrophotometer against a reagent blank. Pipette out 1, 2, 3, 5, and 7 ml of the indole working standard into a series of volumetric flasks. Add 10 ml of Eldrich's reagent, shake well, and separate the acid layer. Read the color absorbance at 570 nm. Draw the standard graph by plotting the concentration of the indole (mg) in the X-axis and the optical density (OD) on the Y-axis. Compute the concentration of indole in the sample extract from the standard graph.

Calculation

$$\text{Indole (mg / 100g)} = \frac{C \times D.F \times 100}{W}$$

where, C - conc. of indole, mg; D.F - dilution factor; W - weight of sample, g; 100 - % conversion

CHAPTER 9

MICROBIOLOGICAL QUALITY AND SAFETY ANALYSIS

Fish and shellfish harbor a wide variety of microorganisms, and those responsible for their spoilage are mainly gram-negative bacteria. Numbers and types of micro-organisms present in aquatic food products are indicators of safety and quality. Microbiological analysis is part of the food safety program and HACCP system in fish processing establishments. Microbiological analysis of fish and shellfish carried out by the following tests.

- Aerobic plate count (APC) of total viable counts (TVC) as colony forming units (CFU)
- Specific spoilage bacteria such as *Pseudomonas*, H_2S producing bacteria
- Pathogens such as *salmonella, enteropathogenic, vibrio, listeria monocytogenes*
- Indicator organisms such as coliforms, enterococci, staphylococci
- Specific organisms associated with processed fishery products

9.1. AEROBIC PLATE COUNT

Aerobic plate count (APC) or total viable count (TVC) or total plate count (TPC) or standard plate count (SPC) provides an estimate of the total number of aerobic microorganisms in foods (AOAC 966.23). APC of fish products indicates quality, shelf life, and post-heat-processing contamination, rather than food safety hazards. Fresh fish often have an APC of 10^4-10^5CFU/g, although there are some tropical fish with an APC of 10^6-10^8 CFU/g without any objectionable quality changes. Plating medium affects the isolation of numbers and types of bacteria because of differences in nutrient and salt requirements of various microorganisms. Incubation temperature effect on the aerobic plate count. APC enumeration includes mixing a series of dilution of fish homogenate, plating onto a non-selective agar medium, and incubating it at $35\pm1°C$ for 48 h.

The microbiological standards for fishery products set out in Commission Regulation (EC) No 2073/2005 of 15 November 2005 prescribes the microbiological criteria for mesophilic aerobic bacteria in different fishery products.

Food Category	Sampling plan		Limits		Analytical reference method
	n	c	m	M	
Cooked Crustacean and Molluscan Shellfish	5	2	10^4 cfu/g	10^5 cfu/g	ISO 4833-1
Shelled and Shucked Products of Cooked Crustacean and Molluscan Shellfish	5	2	5×10^4 cfu/g	5×10^5 cfu/g	ISO 4833-1
Crab meat	5	2	10^5 cfu/g	10^6 cfu/g	ISO 4833-1

where n = number of units comprising the sample; c = number of sample units giving values over m or between m and M; m = minimum level acceptable; M = maximum level tolerable

Food Safety and Standards (Food Products Standards and Food Additives) Third Amendment Regulations 2017 gives the microbiological requirements with respect aerobic plate count for various fishery products. The test method prescribed is IS: 5402/ISO 4833.

Food Category	Sampling plan		Limits (cfu/g)		Stage where criterion applies
	n	c	m	M	
Chilled/frozen finfish	5	3	5×10^5	1×10^7	After chilling/ freezing
Chilled/ frozen crustaceans	5	3	1×10^5	1×10^7	After chilling/ freezing
Chilled cephalopods, bivalves	5	2	1×10^5	1×10^6	After chilling/ freezing
Frozen cooked crustaceans/ frozen heat shucked molluscs	5	2	1×10^5	1×10^6	End of production process

[Table Contd.

Contd. Table]

Food Category	Sampling plan		Limits		Stage where criterion applies
	n	c	m	M	
Dried/salted and dried, and smoked fishery products	5	0	1×10^5		End of production process
Thermally processed fishery products	Commercial sterility				End of production process
Accelerated freeze dried fishery products	5	0		1×10^4	End of production process
Fish mince / surimi and analogues	5	2	1×10^5	1×10^6	End of production process
Fish pickle	5	0		1×10^3	End of production process
Battered and breaded fishery products	5	2	1×10^5	1×10^7	End of production process
Convenience fishery products	5	2	1×10^3	1×10^4	End of production process
Powdered fish based products	5	2	1×10^4	1×10^5	End of production process

where n = number of units comprising the sample; c = number of sample units giving values over m or between m and M; m = minimum level acceptable; M = maximum level tolerable

9.1.1. Direct Plating Method

Determination of APC of foods developed by the Association of Official Analytical Chemists (AOAC), the American Public Health Association (APHA), Food and Drugs Administration (FDA) and International Standard Organization (ISO). The method ISO 4833 – 1&2 2013 specifies a horizontal method for enumeration of microorganisms that are able to grow and form colonies in a solid medium after aerobic incubation at 30°C. Plate Count Agar (PCA) is used for the enumeration of bacteria in water, fish and fishery products. PCA is free from selective supplements and relatively rich in nutrients. It is ideal for the enumeration of viable organisms, either following a pour plate method, or a spread plate method.

Reagents

1. Butterfield's buffered phosphate diluent or 0.85% NaCl
2. Plate count agar (PCA) or Tryptic soy agar (TSA)

   For isolation of marine bacteria, add 1% to 3% NaCl.

   For isolation of bacteria from salted fish, add 5% NaCl

Receipt of samples

Examine the samples immediately upon receipt. If not, freeze them at -20°C. If unfrozen perishables, refrigerate them at 0-4°C not longer than 36 h. If non-perishables like canned or low-moisture foods, store them at room temperature until analysis.

Procedure

Clean the immediate and surrounding work areas with the commercial germicidal agent before analysis. Thaw the frozen samples preferably at 2-5°C within 18 h or at <45°C for not more than 15 min. Take around 50 g of fish sample and add 450 ml of buffered phosphate diluent and blend for 2 min in a homogenizer to get a dilution of 10^{-1}. Prepare decimal dilutions of 10^{-2}, 10^{-3}, 10^{-4}, etc. by transferring 10 ml of the previous dilution to 90 ml of diluent using a sterile pipette within 15 min.

Pour plate technique

In this technique, transfer 1.0 ml of each dilution to duplicate petriplates and then add 12-15 ml of plate count agar or tryptic soy agar (cooled to 45±1°C) to each plate. Mix the sample solution immediately with agar medium thoroughly and uniformly by alternate rotation and back-and-forth motion of plates on a flat level surface. Allow the agar to solidify. Invert the solidified petriplates and incubate for 48 ± 2 h at 35°C.

Spread plate technique

In this technique, prepare plate count agar or tryptic soy agar plates by pouring 12-15 ml of the respective medium in sterile petriplates. Then dry the plates in an incubator set at 45°C in the inverted position for 45 min. From that, pipette out 0.1 ml of each dilution into the pre-poured petriplates and spread using a sterile spreader by alternate rotation and back and forth motion on a flat level surface. Invert the petriplates and incubate for 48 ± 2 h at 35°C. Count and calculate the colonies developed on the agar plates. It is advisable to choose dilutions, which give colonies within a range of 25 – 250. Discard the plates having crowded or spreading colonies.

9.1.2. Guidelines for calculation and report of APCs

1. *Normal plates (25-250):* Count all colony forming units (CFU), including those of pinpointed size. Record the dilution (s) used and count the total number of colonies.

2. *Plates with more than 250 colonies*: When number of CFU / plate exceeds 250 for all dilutions, record the counts as too numerous to count (TNTC). For the plate closest to 250, count the CFU in those portions of plate and report as EAPC.

3. *Spreaders* - Spreading colonies are usually of 3 distinct types:

    a. A chain of colonies

    b. One that develops in film of water between agar and bottom of plate; and

    c. One that forms in film of water at edge or on surface of agar.

        If the plates prepared from sample have excessive spreader growth, report the plates as spreaders. When it is necessary to count plates containing spreaders not eliminated by (a) or (b) above, count each of the 3 distinct spreader types as one source. For the first type, if only one chain exists, count it as a single colony. Compute the spreader count and count the colony to compute the APC.

4. *Plates with no CFU:* When plates from all dilutions have no colonies, report the APC as < 1 times the corresponding lowest dilution used and it is marked as EAPC.

5. *Laboratory accident:* When plate(s) from a sample are known to be contaminated or otherwise unsatisfactory, record the result(s) as laboratory accident (LA).

Computing and recording counts

The results are reported only as the first two significant digits.

i. **Plates with 25-250 CFU**

    Calculate the APC as follows:

$$N = \frac{\Sigma C}{\left[(1 \times n_1) + (0.1 \times n_2) \times d \right]}$$

where, N = number of colonies per ml or g of product; C = sum of all colonies on all plates counted; n_1 = number of plates in first dilution counted; n_2 = number of plates in second dilution counted; d = dilution from which the first counts are obtained.

For Example

1:100	1:1000
232, 244	33, 28

$$N = \frac{(232+244+33+28)}{[(1 \times 2) + (0.1 \times 2)] \times 10^{-2}}$$

$$N = \frac{537}{0.022}$$

$$N = 24,409 = 24,000 = 2.4 \times 10^4$$

ii. All plates with fewer than 25 CFU

When plates from both the dilutions yield fewer than 25 CFU each, record the actual plate count but record the count as < 25 x 1/d, when d is the dilution factor for the dilution from which the first counts are obtained.

For Example

Colonies		EAPC/ml (g)
1:100	1:1000	
18 2		< 2500
0 0		< 2500

EAPC, estimated aerobic plate count

iii. All plates with more than 250 CFU

When plates from both the two dilutions yield more than 250 CFU each (but fewer than $100/cm^2$), estimate the aerobic counts from the plates (EAPC) to nearest 250 and multiply by the dilution.

For Example

Colonies		EAPC/ml (g)
1:100	1:1000	
TNTC	540	640,000

TNTC, too numerous to count; EAPC, estimated aerobic plate count.

iv. All plates with spreaders and/or laboratory accident

Report the plates either as Spreader (SPR) or Laboratory Accident (LA)

v. **All plates with more than an average of 100 CFU/cm^2**

Estimate the APC as greater than 100 times the highest dilution plated, times the area of the plate. Examples below have an average count of 110/cm^2.

For Example

Colonies		EAPC[a]/ml (g)
1:100	1:1000	
TNTC	7,150[b]	> 6500,000
TNTC	6,490[c]	> 5900,000

[a]Estimated aerobic plate count.; [b]Based on plate area of 65 cm^2; [c]Based on plate area of 59 cm^2

Note:

i. An incubation temperature of 20°C yields approximately 10-fold higher counts than at 35°C

ii. To obtain absolute maximum counts from stale fish, incubate the plates at 3°C for 6 days, for the development of obligately psychrotrophic organisms (mostly Vibrios) that do not grow above 15°C.

iii. To get true maximum counts, use the pre-poured and solidified agar plates, and follow the spread plate technique.

9.2. SPECIFIC SPOILAGE ORGANISMS

Shelf-life of aquatic food is correlated with the level of specific spoilage bacteria and not with the APC. The number of specific spoilage microorganisms (SSOs) and the concentration of their metabolites are used as objective quality indicators to determine the shelf-life of aquatic food. It is possible to predict the shelf-life of aquatic food based on knowledge of initial numbers and growth of SSO. The mathematics models are well established for the growth of spoilage bacteria such as *Photobacterium phosphoreum, Shewanella putrefaciens, Brochothrix thermosphacta, Listeria monocytogenes,* and *Clostridium perfringens.* SSO correlates well with the shelf-life of the product. SSO correlates better than the classical total aerobic plate count.

9.2.1. Sulfide producing bacteria

Spoilage is organoleptically detectable when the number of sulfide producing bacteria exceeds 10^{7}cfu/g of the sample. The production of hydrogen sulfide by the microorganism is a common cause of spoilage in aquatic food. The media used to detect the H_2S producing bacteria is the Peptone Iron Agar (PIA). Black/ grey colonies appear due to the precipitation of sulfide complexes formed when H_2S produced from thiosulfate reacts with metal ions such as Fe^{2+} and Pb^{2+}.

Reagents

1. Diluent: Butterfield's buffered phosphate diluent or physiological saline, 0.85% NaCl
2. Peptone Iron agar (PIA)

Procedure

Approximately weigh 25 g of fish aseptically and homogenize well with 225 ml of diluent in a homogenizer or stomacher, to give 10^{-1} dilution. Prepare serial decimal dilutions of 10^{-2}, 10^{-3}, 10^{-4}, and so on depending on the type of spoilage using 9 ml of diluents. Prepare the peptone iron agar plates by pouring 12-15 ml of the medium in sterile petriplates. Place 0.1 ml of each dilution into the pre-poured PIA plates and spread the samples by the spreader. Invert the petriplates and incubate at 35°C for 24-48h. Colonies appear black on PIA plates. Count those colonies within the range of 25 – 250 and calculate the total count.

Calculation

$$\text{Sulfide producing bacteria (CFU/g)} = \frac{\text{No. of colonies} \times \text{dilution} \times 10}{\text{Wt. of the sample, g}}$$

9.2.2. Pseudomonads

Among the psychrotrophic aquatic food spoilage bacteria, relatively few genera are intense spoilage organisms. They are mainly *Pseudomonas*, *Achromobacter,* and *Shewanella*. All pseudomonads are selectively enumerated from aquatic food using Pseudomonas Isolation Agar. Culture media designated for the detection of fluorescent pseudomonads are high in magnesium and low in iron. Fluorescent pseudomonad agar is ideal for the selective isolation of only fluorescent pseudomonads (Sands and Rovira, 1970). Non-selective medium, Pseudomonas Agar F, distinguishes fluorescent *Pseudomonas* colonies from non-fluorescent colonies in fish tissue.

Reagents

1. Diluent: Butterfield's buffered phosphate diluent or physiological saline, 0.85% NaCl

2. Pseudomonas Isolation Agar or Fluorescent Pseudomonad Agar or Pseudomonas Agar F

Procedure

Approximately weigh 25 g of fish aseptically and homogenize with 225 ml of diluent in a homogenizer or stomacher to give 10^{-1} dilution. Prepare the serial decimal dilutions of 10^{-2}, 10^{-3}, 10^{-4}, and so on depending on the type of spoilage using 9 ml of diluents. Prepare the Pseudomonas isolation agar or Fluorescent pseudomonad agar or Pseudomonas agar F plates by pouring 12-15 ml of the medium in sterile petriplates. From that, transfer 0.1 ml of each dilution into the pre-poured respective agar plates and spread following spread plate technique. Invert the petriplates and incubate at 35°C for 24-48h. Count the colonies within the range of 25-250 and calculate the total pseudomonad count.

Calculation

$$\text{Pseudomonads (CFU/g)} = \frac{\text{No. of colonies} \times \text{dilution} \times 10}{\text{Wt. of the sample, g}}$$

9.2.3. Molecular methods for total bacteria

Universal primers used for the quantification of the total bacterial population in fish or shellfish tissue (Lee and Levin, 2006a&b, 2007)

Forward primer: DG74 5' – AGG-AGG-TGA-TCC-AAC-CGA-A-3'

Reverse primer: RW01 5' – ACC-TGG-AGG-AAG-GTG-GGG-AT-3'

Universal primers amplify a 370 bp sequence of the 16S rRNA gene derived from all bacteria (Greisen *et al.*, 1994). An extremely close linear relationship exists between the number of CFU/g determined from plate counts and the total number of genomic targets determined by conventional and real-time polymerase chain reaction. The use of PCR methodology also distinguishes the total number of dead and viable bacteria on fish tissue by using selectively permeable DNA binding dye ethidium bromide monoazide. The amplification of a 207 bp amplicon derived to detect meat spoilage from the 23S rDNA sequence has been done by a pair of primers developed by Venkitanarayanan *et al.* (1996). This assay is also applicable for the collective quantitative PCR enumeration of most of the spoilage bacteria on aquatic foodo.

Forward primer: PF – 5' – AAG-CTT-GCT-GGC-ATC-AGA-AGT-AGT-GC-3'

Reverse primer: PR – 5' – CTC-CGC-CCC-TCC-ATC-GCA-GT-3'

9.3. BACTERIAL PATHOGENS

Aquatic food is associated with the number of bacterial gastroenteritis during the last decades due to the rapid globalization of the food market, an increase of personal and food transportation, and profound changes in food consumption habits. Bacterial pathogens associated with aquatic food are

1. Enteropathogenic *Vibrio*
2. *Listeria monocytogenes*
3. *Salmonella*

Classical microbiological methods can also detect the presence of pathogenic bacteria. The method employs enrichment and isolation of presumptive colonies of bacteria on solid media, followed by the final confirmation of bacteria by biochemical and serological methods. The method is laborious, time-consuming, and requires more than 3-5 days to get accurate results. The use of molecular-based methods is an alternative approach. The advantages of the molecular method include short time, excellent detection limits, specificity, and potential for automation.

9.3.1. Detection of *Vibrio* spp.

Vibrio includes gram-negative, rod or curved rod-shaped oxidase-positive, and facultative anaerobes. Many vibrios are pathogenic to humans implicated in food-borne diseases. Vibrios that grow in media with added sodium chloride are referred to as halophiles, exceot *V. cholerae* and *V. mimicus*. *Vibrio parahaemolyticus, V. vulnificus,* and *V. cholerae* are associated with the consumption of raw or undercooked aquatic food products.

9.3.1.1. *Vibrio parahaemolyticus*

V. parahaemolyticus is one of the important bacterial pathogens associated with aquatic food. It is a gram-negative, motile, facultative, anaerobic rod, halophilic, and non-spore forming bacterium. It lives in estuarine waters and animals. In 1950s, the cause of foodborne illness in Japan was identified as *Vibrio parahaemolyticus*. It accounts for 20-30% of foodborne illness in Japan. Sakazaki proposed the name *V. parahaemolyticus,* for an enteropathogenic halophilic *Vibrio.* It was isolated from many species of fish and shellfish. *V. parahaemolyticus* implicated in numerous outbreaks of food-borne gastroenteritis in the United States and Japan are due to the consumption of raw or insufficiently heated or properly cooked but contaminated aquatic food. Foods frequently involved in outbreaks are fish, oysters, crabs, shrimps, and lobsters cooked before consumption. Heating above 60°C for 15 min destroys this organism. *V. parahaemolyticus* produces heat-stable Kanagawa hemolysin, and some cooking procedures did not destroy hemolysin. Freezing is ineffective in killing the bacteria. Food poisoning is due to gastro-enteritis that develops within 10-15 h. Common symptoms include vomiting, nausea, diarrhea with abdominal cramps, headache, and low-grade fever. Duration of illness is often short that last for 4-6 h and at times longer for 48 h.

Most widely used methods for the detection of *V. parahaemolyticus* in foods are International Organization for Standard (ISO) 8914:2017 (ISO, 2017) and most probable number (MPN) method described in the US Food and Drug Administration (FDA) Bacteriological Analytical Manual (BAM) (Kaysner and Depaola, 2004). A new ISO standard (ISO/TS 21892-1:2017) published is the horizontal method for their detection in food (ISO, 2017). The microbiological standards for fishery products set out in Commission Regulation (EC) No 2073/ 2005 of 15 November 2005 gives the microbiological criteria for cooked crustacean and molluscan shellfish.

Food Category	Sampling plan		Limits		Analytical reference method
	n	c	m	M	
Vibrio parahamolyticus	5	2	10 cfu/g	100 cfu/g	ISO 8914(MPN) ISO 21872-1:2017

where n = number of units comprising the sample; c = number of sample units giving values over m or between m and M; m = minimum level acceptable; M = maximum level tolerable

9.3.1.1.1. ISO Standard Method

ISO 21872-1:2017 specifies a horizontal method for the detection of enteropathogenic *Vibrio* spp. The species detectable include *Vibrio parahaemolyticus, Vibrio cholerae* and *Vibrio vulnificus.*

Reagents

1. Phosphate buffered saline (PBS)
2. Salt polymyxin B broth

3. Alkaline peptone saline broth or saline glucose culture medium with SDS

4. TCBS agar

5. Triphenyl tetrazolium chloride tryptone agar

   Above media and diluents contain 3% NaCl.

Procedure

Approximately weigh 10 g of the sample and transfer into 90 ml of two enrichment broths viz. i. salt polymyxin B broth and ii. alkaline saline peptone water or saline glucose culture medium supplemented with sodium dodecyl sulfate and incubate at 41°C for 18 h. Streak a loopful of inoculum onto two selective media viz. i. thiosulfate citrate bile salt sucrose agar (TCBS) and ii. triphenyl tetrazolium chloride soya tryptone agar (TSAT). Incubate the TCBS agar plates for 18 h or TSAT agar plates for 20-24 h at 35°C. The presumptive colonies of *V. parahaemolyticus* on TCBS agar plates appear as 2-3 mm, smooth and green colonies or on TSAT agar plates as 2-3 mm, smooth flat, and dark red colonies. Confirm the colonies by subjecting them to various biochemical tests.

9.3.1.1.2. FDA-BAM method

Reagents

1. Phosphate buffered saline (PBS)

2. Alkaline Peptone Salt (APS) broth

3. TCBS agar

4. Trypticase soy agar (TSA)

   Above media and diluents contain 3% NaCl.

Procedure

9.3.1.1.2.a. MPN method

Approximately weigh 50 g of the sample, blend in a homogenizer and transfer to 450 ml of sterile phosphate buffered saline (PBS). Prepare a ten-fold dilution with the same diluent. Inoculate 1 ml portions of each dilution into sets of 3 tubes containing 10 ml alkaline peptone salt (APS) broth and incubate the tubes at 35-37°C for 16-18 h. Complete the inoculations in MPN tubes within 15-20 min of dilution preparation.

9.3.1.1.2.b. Direct plating method

Alternatively, transfer 0.1 ml of the inoculum from each dilution onto TCBS agar plates. Spread the inoculum over the surface of agar plate using sterile bent glass streaking rod. Retain the plates in upright position until inoculum is absorbed by agar. Invert the plates and incubate for 24 h at 35°C. Observe the plates for typical colonies of *V. parahaemolyticus,* which appear as round, 2-3 mm in diameter, green or blue-green colonies. Count the positive colonies, calculate and express the results as number of CFU per gram. Pick up three or five typical colonies from TCBS plates and sub-culture on gelatin salt (GS) agar or trypticase soy agar (TSA) plates for purification of the colonies for further biochemical identification.

Biochemical confirmation of *V. parahaemolyticus*

No	Tests	Reaction
1.	Gram stain	Gram-negative asporogenous rod
2.	TSI	Alkaline slant/acid butt, gas production-negative, H_2S-negative
3.	Hugh-Leifson test	Glucose oxidation and fermentation-positive
4.	Cytochrome oxidase	Positive
5.	Arginine dihydrolase test	Negative
6.	Lysine decarboxylase test	Positive
7.	Voges-Proskauer test	Negative
8.	Growth at 42ºC	Positive
9.	Halophilism test	0% NaCl-negative; 3, 6, and 8% NaCl-positive; 10% NaCl-negative or poor
10.	Sucrose fermentation	Negative
11.	ONPG test	Positive
12.	Arabinose fermentation	Usually positive (variable)
13.	O/129 sensitivity	Sensitive to 150 µg, resistant to 10 µg

9.3.1.1.3. PCR method

Several PCR methods are available to detect *V. parahaemolyticus* in aquatic food products and related environments. Gene marker used for *V. parahaemolyticus* specific detection is the thermolabile hemolysin (*tlh*) gene (Wang and Levin, 2004). PCR methods are also available for the detection of *V. parahaemolyticus* by targeting toxin transcriptional activator (*toxR*) gene (Kim *et al.,* 1999), B subunit of DNA gyrase (*gyrB*) gene (Venkateswaran *et al.,* 1998), thermostable direct hemolysis (*tdh*), and tdh-related hemolysin (*trh*) genes (Tada *et al.*, 1992). A multiplex PCR method is available for the detection of the total and pathogenic strains of *Vibrio* by targeting *tlh, tdh,* and *trh* genes (Nordstrom *et al.*, 2007). The emergence of the O3:K6 serotype and its widespread distribution have led to the development of a PCR method for specific detection of the open reading frame 8 of phage f237 (*orf8).*

9.3.1.2. *Vibrio vulnificus*

V. vulnificus produces one of the most severe foodborne infections with a fatality rate greater than 50%. It causes fatal septicemia, wound infections, and gastroenteritis, especially in immune-compromised individuals. It was first isolated by the Center for Disease Control in 1964 and found worldwide. Most of the outbreaks in the US are reported during the summer season.

Detection protocol approved by the FDA BAM method for *V. vulnificus* is the MPN enrichment in APW coupled with isolation in selective medium and biochemical/molecular confirmation. Another method is by the direct isolation on minimally selective media followed by the identification of *V. vulnificus* by colony blot DNA-DNA hybridization. Recently, ISO/TS 21872-1:2007 standard published a horizontal method for the detection of other pathogenic *Vibrio* spp. (ISO, 2007). Current available selective media for *V. vulnificus* are TCBS, Vibrio vulnificus agar, SDS polymyxin agar, cellobiose polymyxin B colistin agar, cellobiose colistin agar, Vibrio vulnificus enumeration (VVE) agar, and Vibrio vulnificus media (VVM).

9.3.1.2.1. PCR method

A PCR method based on the cytolysin gene (*vvhA*) developed by Hill *et al.* (1991) is used for the detection of *V. vulnificus.* Kumar *et al.* (2006) developed another PCR method based on the *gyrB* gene. A real-time PCR method reported by Vickery *et al.* (2007) classifies V. *vulnificus* based on the 16S rRNA genotype.

9.3.1.3. *Vibrio cholerae*

V. cholerae was first described as the cause of cholera by Pacini in 1854. It is a short, comma-shaped curved, cylindrical rod, with rounded or slightly pointed ends. Cholera is a food and waterborne disease caused in humans when the organism enters the intestine through contaminated food or water. Cholera causes profuse watery diarrhea and vomiting that leads to shock and death in less than 24 h in severe cases. Pathogenic *V. cholerae* produces a heat-sensitive enterotoxin. *V. cholerae* comprises several somatic (O) antigen groups, including

O group 1, which is associated with classical and El Tor biotypes. Serotypes of *V. cholerae* O1 include Inaba, Ogawa, and Hikojima. *V. cholerae* non-O1 (referred to as non-agglutinable or NAG vibrios) also causes gastrointestinal disease. Serotype O139 is an exception as it also produces classic cholera symptoms. It caused a new epidemic of cholera in India and Bangladesh and first identified in 1992. Non-O1 *V. cholerae* is more in estuarine waters and aquatic food in the United States than O1 and O139 serogroups. *V. cholerae* can also grow in media lacking sodium chloride. Testing of *V. cholerae* O1 and non-O1 isolates is recommended for the production of cholera toxin. FDA-BAM method for detection of *V. cholerae* in food relies on the overnight enrichment in APW, followed by isolation on selective medium and final biochemical and molecular confirmation (Kaysner and DePaola, 2000).

Food Safety and Standards (Food Products Standards and Food Additives) Third Amendment Regulations 2017 gives the microbiological requirements for various fishery products for *Vibrio cholerae* (O1 and non O1). The test method prescribed is as per BAM Chapter 9 USFDA, May 2004.

Food Category	Sampling plan		Limits	
	n	c	m	M
Chilled/frozen finfish, crustaceans, cephalopods and bivalves	5	0	Absent in 25 g	
Frozen cooked crustaceans/ frozen heat shucked molluscs	5	0	Absent in 25 g	
Smoked fishery products	5	0	Absent in 25 g	
Accelerated freeze dried fishery products	5	0	Absent in 25 g	
Fish mince / surimi and analogues	5	0	Absent in 25 g	
Battered and breaded fishery products	5	0	Absent in 25 g	
Convenience fishery products	5	0	Absent in 25 g	

where n = number of units comprising the sample; c = number of sample units giving values over m or between m and M; m = minimum level acceptable; M = maximum level tolerable

9.3.1.3.1 FDA-BAM method

Reagents

1. Alkaline Peptone Water (APW)
2. Buffered phosphate saline (PBS)
3. Thiosulphate-citrate-Bile Salt-Sucrose (TCBS) Agar

Procedure

Aseptically weigh 25 g of the sample, blend in a homogenizer for 2 min and transfer to 225 ml of Alkaline Peptone Water (APW). Incubate the flask at 35 ± 2°C for primary enrichment. Subsequently, streak a loopful of inoculum from APW onto TCBS agar plates after 6-8 h and after 18-24 h of enrichment and incubate for 24 h at 35 ± 2°C. Typical *V. cholerae* colonies on TCBS Agar plates appear as large, smooth, yellow and slightly flattened with an opaque center and translucent peripheries. Purify three or more typical colonies from each of the TCBS plates by streaking onto trypticase soy agar plates. Perform the following tests to ensure the biochemical identity of *V. cholerae*.

Biochemical confirmation of *Vibrio cholerae*

No	Tests	Reaction
1.	Gram stain	Gram-negative asporogenous rod or curved rod
2.	TSI or KIA appearance	Acid slant/acid butt, gas production negative, H_2S-negative
3.	Hugh-Leifson test	Glucose fermentation and oxidation-positive
4.	Cytochrome oxidase	Positive
5.	Arginine dihydrolase test	Negative
6.	Lysine decarboxylase test	Positive
7.	Voges-Proskauer test	El Tor biotype-positive, classical biotype-negative, *V. mimicus* negative
8.	Growth at 42°C	Positive
9.	Salt tolerance	% 0 NaCl-positive; 3% NaCl-positive; 6% NaCl-usually negative. Some strains of *V. cholerae* non-O1 may not grow in 0% NaCl.
10.	Sucrose fermentation	Positive (negative for *V. mimicus*)
11.	ONPG test	Positive
12.	Arabinose fermentation	Negative
13.	O/129 sensitivity	Sensitive to 10 and 150 µg O/129

9.3.1.3.2 PCR method

A PCR method targeting the cholera toxin operon, *ctxAB* is developed by Koch *et al.* (1993) with 100% selectivity. A real-time PCR method developed for the detection of toxigenic *V. cholerae* in seafood by Blackstone *et al.* (2007) targets the *ctxA* gene.

A multiplex PCR assay is developed for the simultaneous detection of total and toxigenic *Vibrio* spp. Total vibrios is targeted using genus-specific RNA polymerase subunit A (*rpoA*) gene and toxin-producing *Vibrio cholerae* strains using two sets of primers viz. cholera toxin subunit A (*ctxA*) and repeat in toxin subunit A (*rtxA*) genes. The primer sets for *rpoA* gene is F - AAA TCA GGC TCG GGC CCT and R- GCA ATT TT(A/G)TC(A/G/T)AC(C/T)GG with product size of 242 bp; for *ctxA* gene F - ACA GAG TGA GTA CTT TGA CC and R- ATA CCA TCC ATA TAT TTG GGA with product size of 308 bp, and for *rtxA* gene F- CTG AAT ATG AGT GGG TGA CTT ACG and R GTG TAT TGT TCG ATA TCC GCT ACG with a product size of 417 bp (Jeyasekaran *et al.*, 2011).

9.3.2. Detection of *Listeria monocytogenes*

Listeria monocytogenes is a gram-positive, non-spore forming, catalase-positive, microaerophilic rod-shaped bacteria. It is motile at room temperature, and non-motile at 37°C. *Listeria* includes 6 species viz., *L. monocytogenes*, *L. ivanovii*, *L. innocua*, *L. welshimeri*, *L. seeligeri*, and *L. grayi*. Of which, *L. ivanovii* and *L. monocytogenes* are pathogenic for mice, but only *L. monocytogenes* is associated with human illness, referred to as listeriosis. Victims of severe listeriosis are usually immunocompromised, and those at the highest risk are cancer patients, individuals taking drugs, alcoholics, pregnant women, persons with low stomach acidity, and AIDS patients. Severe listeriosis causes meningitis, abortions, and septicemia and can lead to death. Apathogenic *L. innocua* is most frequently isolated from aquatic food. All virulent strains produce β-hemolysis on blood agar and produce a toxin called listeriolysin. The greatest threat of listeriosis is ready-to-eat products. It is more prevalent in raw fish, cooked crabs, raw and cooked shrimp, raw lobster, surimi, and smoked fish. Thorough cooking prevents hazards associated with *L. monocytogenes*. *L. monocytogenes* should be absent in fish and fishery products. The most prevalent serotype is 1/2a. There are also other serotypes such as 1/2b, 1/2c, and 4b.

The microbiological standards for fishery products set out in Commission Regulation (EC) No 2073/2005 of 15 November 2005 gives the microbiological criteria for *L. monocytogenes* in ready-to-fishery products.

Food Category	Sampling plan		Limits		Analytical reference method	Stage where the criterion applies
	n	c	m	M		
RTE foods able to support growth of *L. monocytogenes*	5	0	100 cfu/g		EN/ ISO 11290-2	Products in market during their shelf-life
	5	0	Absence in 25g		EN/ ISO 11290-1	Under the control of FBO
RTE foods unable to support growth of *L. monocytogenes*	5	0	100 cfu/g		EN/ ISO 11290-2	Products in market during their shelf-life

where n = number of units comprising the sample; c = number of sample units giving values over m or between m and M; m = minimum level acceptable; M = maximum level tolerable

Food Safety and Standards (Food Products Standards and Food Additives) Third Amendment Regulations 2017 gives the microbiological requirements for various fishery products for *Listeria monocytogenes*. The test method prescribed is IS:14988 Part 1 & 2/ ISO 11290-1 & 2.

Food Category	Sampling plan		Limits	
	n	c	m	M
Frozen cooked crustaceans/ frozen heat shucked molluscs	5	0	Absent in 25 g	
Smoked fishery products	5	0	Absent in 25 g	
Accelerated freeze dried fishery products	5	0	Absent in 25 g	
Fish mince / surimi and analogues	5	0	Absent in 25 g	
Battered and breaded fishery products	5	0	Absent in 25 g	
Convenience fishery products	5	0	Absent in 25 g	

where n = number of units comprising the sample; c = number of sample units giving values over m or between m and M; m = minimum level acceptable; M = maximum level tolerable

9.3.2.1 ISO Standard method

The ISO has developed reference methods for the detection and enumeration of *L. monocytogenes* vide ISO 11290-1 and 11290-2 (ISO, 2017). Listeria spp. can be isolated using two selective media such as Agar *Listeria* of Ottaviani and Agosti (ALOA™) or Oxoid Chromogenic *Listeria* agar (OCLA) and PALCAM or Oxford media. The ALOA allows diagnosis of *Listeria monocytogenes* based on two enzyme activities : β-glucosidase that hydrolyzes a chromogenic compound 5-bromo-4- chloro-3-indolyl-β-D glucopyranoside, and phosphatidylinositol-specific phospholipase C (Pl-PLC) that hydrolyze the purified substrate, phosphatidylinositol. OCLA identifies *Listeria* spp. as they cleave chromogen X-glucoside by ß-glucosidase, while lithium chloride, polymyxin B and nalidixic acid inhibits enterococci that possess this enzyme, and differentiates *Listeria monocytogenes* and pathogenic *Listeria ivanovii* by their ability to hydrolyze phosphotidylinositol or lethicin. The PALCAM allows the diagnosis of *Listeria* spp., as they hydrolyze aesculin. The inclusion of a ferric salt in the medium leads to the formation of a black complex by Listeria. *L. monocytogenes* does not ferment mannitol. The lithium chloride and ceftazidime suppress *Enterococcus* spp. and an occasional strain of *Micrococcus*. The ceftazidime inhibits the other commensals present in fresh foods. Oxford media utilizes selective inhibitory components, lithium chloride, acriflavin, colistin sulphate, cefotetan, cycloheximide or amphotericin B and fosfomycin, and also indicators, aesculin and ferrous iron to differentiate *Listeria monocytogenes*.

Reagents

1. Primary enrichment broth (Half Fraser broth)
2. Secondary enrichment broth (Fraser Broth)
3. Agar Listeria according to Ottaviani and Agosti (ALOA)
4. Oxford agar (OXA)
5. PALCAM (Polymyxin-Acriflavine-Lithium chloride–Ceftazidime–Aesculin–Mannitol) agar

Procedure

Homogenize 25 g of the sample in 225 ml of primary enrichment medium (Half Fraser broth) and incubate at 30°C for 24h. Transfer 0.1ml of primary culture to 10ml of Fraser Broth and incubate for 48 h at 37°C. Streak the primary culture onto Agar Listeria of Ottaviani and Agosti (ALOA™) or Oxoid Chromogenic Listeria agar (OCLA) and PALCAM or Oxford media and incubate at 37°C for 24-48h. Typical colonies of *L. monocytogenes* appear as blue-green coloured colonies with an opaque halo around the colonies in ALOA plates, as blue colonies with or without halos in OCLA plates, as black colonies with sunken center in PALCAM plates, and as black phenolic colonies on Oxford agar. The typical colonies are further confirmed by biochemical tests.

9.3.2.2. FDA-BAM method

Reagents

1. Listeria enrichment broth base (BLEB)
2. Oxford medium (OXA)
3. PALCAM (Polymyxin-Acriflavine-Lithium chloride–Ceftazidime–Aesculin–Mannitol) agar
4. Modified Oxford agar (MOX)
5. Lithium chloride phenylehtanol moxalactam (LPM) agar
6. Tryptic soy agar + 0.6% yeast extract (TSAYE)

Procedure

Take 25 g of sample homogenate and transfer to 225 ml of buffered Listeria enrichment broth base (BLEB) without selective agents. Incubate the flask for 4 h at 30°C and add selective agents such as acriflavine, nalidixic acid, and cycloheximide and again incubate for another 44h at 30°C. Transfer 10 ml of pre-enriched sample to 90 ml of UVM broth and incubate at 37°C for 24 h for primary enrichment. Streak the enriched sample onto Oxford agar, PALCAM, modified Oxford agar (MOX) and Lithium chloride phenylehtanol moxalactam (LPM) agar fortified with esculin and ferric salt. Incubate the plates at 35°C for 24-48 h for Oxford, PALCAM, or MOX plates or at 30°C for 24-48h for fortified LPM plates. The typical colonies of *Listeria* have a black halo on OXA and PALCAM plates. Then, streak five or more typical colonies onto TSAYE plates and incubate at 30°C for 24-48 h for confirmation by biochemical tests (Hitchins, 2003).

Biochemical confirmation of *Listeria monocytogenes*

No	Tests	Reactions	Results
1.	Motility at 30°C	Tumbling motility	Positive
2.	Catalase	Effervescence	Positive
3.	Gram staining	Purple short rod	Positive
4	Nitrate reduction test	No color change	Negative
5.	Dextrose	Acid; gas	Positive
6.	Esculin	Acid	Positive
7.	Maltose	Acid	Positive
8.	Rhamnose	Acid	Positive
9.	Mannitol	No acid	Negative
10.	Xylose	No acid	Negative
11.	ß Hemolysis	Hemolytic	Positive
12.	CAMP reaction with *S. aureus*	Hemolytic	Positive

Differentiation of *Listeria* species

Species	Hemolytic[a]	Nitrate reduction	Acid production from			Virulence in mouse
			Mannitol	Rhamnose	Xylose	
L. monocytogenes	+	-	-	+	-	+
L. ivanovii	+	-	-	-	+	+
L. innocua	-	-	-	V[b]	+	-
L. welshimeri	-	-	-	V[b]	+	-
L. seeligeri	+	-	-	-	+	-
L. grayi[c]	-	V	+	V	-	-

[a]Sheep blood stab; [b]V, variable

[c]*L.grayi* now includes the former nitrate-reducing, rhamnose variable species *L. murrayi*

9.3.2.3. PCR method

Four sets of primers are used to confirm *Listeria* spp. by PCR. One set targeting the 16S rRNA gene is for the identification of *Listeria* spp. Three sets targeting the *hly* and *iap* genes are for the identification of *L. monocytogenes*. Agersborg *et al.* (1997) develops a specific PCR method first time for the detection of *L. monocytogenes* in aquatic food products.

A PCR assay is developed for the detection of *Listeria monocytogenes* published using *hlyA* gene (Park *et al.*, 2012). Forward primer is GCATCTGCAT TCAATAAAGA and reverse primer is TGTCACTGCATCTCCGTGGT with a product size of 174 bp. The PCR assay targeting *hlyA* gene is even used for the confirmation of *L. monocytogenes* isolated from seafood contact surfaces (Selvaganapathi *et al.*, 2018).

9.3.3. Detection of *Salmonella* spp.

Salmonella is a rod-shaped, non-spore-forming, and gram-negative motile bacterium, except *S. gallinarum* and *S. pullorum*. It is a catalase-positive, oxidase-negative, and facultatively anaerobe. *S.* Typhi and *S.* Paratyphi cause septicemia and produce typhoid or typhoid-like fever in humans. Other forms of Salmonellosis produce mild symptoms. Acute symptoms include nausea, vomiting, abdominal cramps, diarrhea, fever, and headache. Chronic consequences include arthritic symptoms that follow 3 - 4 weeks after onset of acute symptoms. The onset of the illness is usually 6 - 48 h. The infective dose is as few as 15–20 cells depending upon age and health of host and strain differences. Of the outbreaks recorded in the World Health Organization (WHO), *Salmonella* is the causative agent in 55% of the cases. Food items causing hazards include raw or undercooked aquatic food products. *Salmonella* hazards can be prevented by heating fish/ shellfish to kill the bacteria (e.g. 24 sec at 74°C), holding chilled aquatic food below 4.4°C, preventing post-cooking cross-contamination, and prohibiting people, who are ill or carriers of *Salmonella* from working in food operations. The most prevalent serotype is *Salmonella enterica* serotype Weltevreden. Other serotypes are *S. enterica* serotype Worthington and *S. enterica* serotype Newport. There are two methods for the detection of *Salmonella* spp.

The microbiological standards for fishery products set out in Commission Regulation (EC) No 2073/2005 of 15 November 2005 gives the microbiological criteria for *Salmonella* in cooked crustaceans and molluscan shellfish, and live bivalve molluscs.

Food Category	Sampling plan		Limits		Analytical reference method	Stage where the criterion applies
	n	c	m	M		
Salmonella	5	0	Absence in 25 g		EN/ ISO 6579	Products in market during their shelf-life

where n = number of units comprising the sample; c = number of sample units giving values over m or between m and M; m = minimum level acceptable; M = maximum level tolerable

Food Safety and Standards (Food Products Standards and Food Additives) Third Amendment Regulations 2017 gives the microbiological requirements for various fishery products for *Salmonella*. The method prescribed is IS: 5887 Part 3 or ISO 6579.

Food Category	Sampling plan		Limits	
	n	c	m	M
Chilled/ Frozen finfish, crustaceans, cephalopods	5	0	Absent in 25 g	
Chilled/ Frozen bivalves	10	0	Absent in 25 g	
Frozen cooked crustaceans/ frozen heat shucked molluscs	5	0	Absent in 25 g	
Dried/ salted and dried fishery products	5	0	Absent in 25 g	
Fermented fishery products	10	0	Absent in 25 g	
Smoked fishery products	5	0	Absent in 25 g	
Accelerated freeze dried fishery products	5	0	Absent in 25 g	
Fish mince / surimi and analogues	5	0	Absent in 25 g	
Fish pickle	5	0	Absent in 25 g	

[Table Contd.

Contd. Table]

Food Category	Sampling plan		Limits	
	n	c	m	M
Battered and breaded fishery products	5	0	Absent in 25 g	
Convenience fishery products	5	0	Absent in 25 g	
Powdered fish based products	5	0	Absent in 25 g	

where n = number of units comprising the sample; c = number of sample units giving values over m or between m and M; m = minimum level acceptable; M = maximum level tolerable

9.3.3.1. ISO standard method

ISO 6579-1:2017 specifies a horizontal method for the detection of *Salmonella*. This horizontal is a method detects most of the *Salmonella* serovars. The selective enrichment medium is a modified semi-solid Rappaport-Vassiliadis (MSRV) medium detecting of motile *Salmonella* and not non-motile *Salmonella* strains. In xylose lysine deoxycholate (XLD) agar, lysine differentiates *Salmonella* as they possess lysine decarboxylase enzyme. Bismuth Sulphite Agar (BSA) differentiates *Salmonella* through their ability to cleave the two chromogens by specific enzymes, caprylate esterase and β-glucosidase. Brilliant Green Agar (BGA) uses brilliant green to inhibit the growth of gram-positive bacteria and most gram-negative bacilli other than *Salmonella* spp.

Reagents

1. Buffered peptone water (BPW)
2. Rappaport-Vassiliadis medium with soya broth (RVS)
3. Muller–Kauggman tetrathionate novobiocin (MKTTn) broth
4. Xylose lysine deoxycholate (XLD) agar
5. Bismuth Sulphite Agar (BSA)
6. Hektoen enteric (HE) agar

Procedure

Aseptically weigh 25 g of the sample, homogenize with buffered peptone water (BPW) and incubate at 37°C for 18h. Subsequently, transfer 0.1 ml of pre-enrichment aliquot into 10 ml of Rapport-Vassiliadis (RV) medium with soya (RVS broth) and incubate for 24 h at 41.5°C. Also, transfer 1 ml of aliquot into 10 ml of Muller–Kauggman tetrathionate novobiocin (MKTTn) broth and incubate for 24 h at 37°C. After the 24 h incubation, streak a loop of the RVS and MKTTn broths onto XLD agar or Bismuth Sulphite Agar or Brilliant Green Agar and incubate the plates at 37°C for 24 h. The *Salmonella* colonies appear as red in XLD agar, as purple or black in BSA, and as red-pink-white surrounded by red zones in BGA. The typical *Salmonella* colonies are further confirmed by biochemical and serological tests.

9.3.3.2. FDA-BAM method

Reagents

1. Lactose Broth (LB)
2. Rappaport-Vassiliadis (RV) medium
3. Selenite Cystine Broth (SCB)
4. Tetrathionate Broth (TTB)
5. Bismuth Sulphite Agar (BSA)
6. Xylose Lysine Deoxycholate (XLD) Agar
7. Hektoen enteric (HE) agar

Procedure

Aseptically weigh 25 g of fish sample, blend for 2 min in a homogenizer and transfer to 225 ml of sterile lactose broth. After 1 h at room temperature, add 2.25 ml of steamed Tergitol anionic 7 or Triton X-100 and incubate the broth at 24 h at 35°C for pre-enrichment. Subsequently, transfer 0.1 ml of pre-enrichment aliquot into 10 ml Rappaport-Vassiliadis (RV) medium or Selenite cysteine (SC) broth and add 1 ml into 10 ml tetrathionate (TT) broth. Incubate the tubes for 24 h at 42 ± 0.2°C in a water bath. Streak a loopful of enriched culture from RV or SCB and TT broth onto Bismuth sulfite (BS) agar, Xylose lysine deoxycholate (XLD) agar, and Hektoen enteric (HE) agar. Incubate the plates for 24 h at 35°C. Examine the plates for typical *Salmonella* colonies based on the morphology. Finally, confirm the typical *Salmonella* colonies by biochemical and serological tests. Bismuth sulphite (BS) agar *Salmonella* appear as brown, grey, or black colonies; sometimes have a metallic sheen. Surrounding medium may turn black with increased incubation, producing 'halo' effect. Xylose lysine deoxycholate (XLD) agar *Salmonella* appear as pink colonies with or without black centers. Hektoen enteric (HE) agar *Salmonella* appear as blue-green to blue colonies with or without black centers.

Biochemical confirmation for *Salmonella*

No	Test or substrate	Result		*Salmonella* species reaction[a]
		Positive	**Negative**	
1.	Glucose (TSI)	Yellow butt	Red butt	+
2.	Lysine decarboxylase (LIA)	Purple butt	Yellow butt	+
3.	H$_2$S (TSI and LIA)	Blackening	No blackening	+
4.	Urease	Purple red	No colour change	-
5.	Lysine decarboxylase broth	Purple colour	Yellow colour	+
6.	Phenol red dulcitol broth	Yellow colour and/ or gas	No gas; No colour change	+[b]
7.	KCN broth	Growth	No growth	-
8.	Malonate broth	Blue colour	No colour change	-
9.	Indole test	Violet colour	Yellow colour	-
10.	Polyvalent flagellar test	Agglutination	No agglutination	+
11.	Polyvalent somatic test	Agglutination	No agglutination	+
12.	Phenol red lactose broth	Yellow colour and/ or gas	No gas; No colour change	-[c]
13.	Phenol red sucrose broth	Yellow colour and/ or gas	No gas; No colour change	-
14.	Voges-Proskauer test	Pink-to-red colour	No colour change	-
15.	Methyl red test	Diffuse red colour	Diffuse yellow colour	+
16.	Simmons citrate	Growth; blue colour	No growth; no colour change	v

[a]+, 90% or more positive in 1 or 2 d; -, 90% or more negative in 1 or 2 d; v, variable.
[b]Majority of *S. arizonae* cultures are positive.
[c]Majority of *S. arizonae* cultures are negative.

9.3.3.3. PCR method

A novel PCR method developed is for the detection of *Salmonella entertidis* targeting the *Salmonella invA* gene (Wang and Yeh, 2002).

A multiplex PCR based assay is developed for multiple confirmation of genus *Salmonella* serovars viz. Typhi, ParatyphiA, Typhimurium, Enteritidis and Weltevreden by targeting *fimA, himA, hns, invA* and *hto* genes giving amplified products of sizes of 85, 123, 152, 275 and 496 bp, respectively. This assay has a sensitivity to detect 5 cells of *Salmonella* and 1,000 fg of genomic DNA

(Jeyasekaran *et al.*, 2012). The primers and product sizes of the target genes are given below:

Target genes	Primers	Product size
fimA	F- CCT TTC TCC ATC GTC CTG AA R- TGG TGT TAT CCG CCT GAC CA	85
himA	F-CGT GCT CTG GAA AAC GGT GAG R-CGT GCT GTA ATA GGA ATA TCT TCA	123
hns	F-TAC CAA AGC TAA ACG CGC AGC T R- TGA TCA GGA AAT CTT CCA GTT GC	152
invA	F- TAT CGC CAC GTT CGG GCA A R- TCG CAC CGT CAA AGG AAC C	275
hto	F- ACT GGC GTT ATC CCT TTC TCT GGT TG R- ATG TTG TCC TGC CCC TGG TAA GAG A	496

A MPCR assay is developed for the detection of *Salmonella enterica* serovars Typhi, Paratyphi A, Typhimurium, Enteritidis, Weltevreden, Bovismorbificians, Brunei, Arizonae and Infantis with *Salmonella* genus specific *himA* gene and an internal amplification control (IAC) region to avoid false negative results from fish/shrimp tissue after 4 h of pre-enrichment. The *himA* gene primers are F-CGT GCT CTG GAA AAC GGT GAG and R-CGT GCT GTA ATA GGA ATA TCT TCA having a product size of 123 bp and the 16S-23S internal transcribed spacer region served IAC primers are F-TAT AGC CCC ATC GTG TAG TCA GAA C and R-TGC GGC TGG ATC ACC TCC TT having a product size of 312 bp. The non-amplification of target DNA and the amplification of non-target IAC indicate that false-negative result is eliminated (Thirumalairaj et al., 2011).

9.3.4. Multiplex PCR for detection of pathogens

A multiplex PCR for the simultaneous detection of *Escherichia coli, S. enterica* serotype Typhimurium, *V. vulnificus, V. cholerae*, and *V. parahaemolyticus* is developed at the University of Alabama (Brasher *et al.*, 1998) targeting *uidA* gene of *E.coli, invA* gene of *S. typhimurium, cth* gene of *V. vulnificus, ctx* gene of *V. cholerae* and *tlh* gene of *V. parahaemolyticus*.

Procedure

Blend 1 g of fish tissue in 30 ml of APW and incubate at 35°C for 6h. After enrichment, centrifuge the homogenate and extract the DNA using the extraction kit. To achieve maximum sensitivity, subject 5 μL aliquot of the initial multiplex PCR amplified products to a re-amplification process by a second PCR. The minimum level of detection of each target in a single multiplex PCR is 100 cfu/g. The detection limit can be improved to 10 cells cfu/g by the second PCR.

9.4. INDICATOR ORGANISMS

Indicator organisms measure the potential fecal contamination of environmental samples. Indicator organisms need not be pathogens. Coliform bacteria in water samples may be quantified using the most probable number (MPN) method. Non-coliform bacteria such as *Streptococcus brevis* and certain clostridia are also indices of fecal contamination.

9.4.1. Enumeration of coliforms, faecal coliforms and *Escherichia coli*

Coliforms are gram-negative, rod-shaped facultatively anaerobic bacteria. The coliform group includes species from the genera *Escherichia*, *Klebsiella*, *Enterobacter*, and *Citrobacter*. They are used as indicator microorganisms to serve as a measure of fecal contamination. Coliform gets easily killed by heat and hence, useful for testing cooked fish and fishery products for post-process contamination. Coliform counts cannot differentiate between fecal and non-fecal contamination, and hence it is better to perform the fecal coliform test. Fecal coliforms are coliforms that ferment the lactose in EC medium with gas production within 48 h at 44.5°C. Fecal coliforms are directly associated with fecal contamination from warm-blooded vertebrates. *E. coli* usually makes up to 75-95% of the fecal coliform count. *E. coli* naturally inhabit the intestinal tracts of all warm-blooded animals, including humans. Most forms of the bacteria are not pathogenic and serve useful functions in the intestine. Pathogenic strains of *E. coli* get transferred to aquatic food through sewage pollution or by contamination after harvest. *E. coli* food infection causes abdominal cramping, watery or bloody diarrhea, fever, nausea, and vomiting. Some *E. coli* strains may be only weakly lactose-positive or even lactose-negative.

Enterovirulent *E. coli* (EEC) strains

Enterohemorrhagic *E. coli* (EHEC) strain causes hemorrhagic colitis and hemolytic uremic syndrome in humans. Six verocytotoxins are identified within this group, but only *stx-1* and *stx-2* seem to be important in human infections. *E. coli* O157:H7 is the principle serotype of this group. Enteroinvasive *E. coli* (EIEC) causes a diarrheal illness similar to shigellosis. Enterotoxigenic *E. coli* (ETEC) causes travelers' diarrhea and infant diarrhea in developing countries. ETEC produces a heat-labile toxin (LT) and/or a heat-stable toxin (ST). Enteropathogenic

E. coli (EPEC) causes infant diarrhea. Enteroadherent *E. coli* (EAEC) is a new type and not yet fully characterized like other types. Hazards from *E. coli* are prevented by heating aquatic food to kill the bacteria, by holding chilled aquatic food below 4.4°C, and by preventing post-cooking cross-contamination. The maximum allowable number of *E. coli* in raw aquatic food products is 20/g, and absent in cooked products. Fecal coliforms in fish processing water should be less than 1/100ml by the MPN technique. Coliform, fecal coliforms, and *E. coli* counts are determined by the MPN method. Total coliforms and *E. coli* counts are also determined by the direct plating techniques.

The microbiological standard for fishery products set out in Commission Regulation (EC) No 2073/2005 of 15 November 2005 gives in the microbiological criteria for *E.coli* in different food.

Food Category	Sampling plan		Limits		Analytical reference method	Stage where the criterion applies
	n	c	m	M		
Shelled and shucked products of cooked crustacean and molluscan shellfish	5	2	1 cfu	10 cfu	EN / ISO 16649-1	End of the production process
Live bivalve molluscs	1	0	<230/100g		EN / ISO 16649-1	Production area

Where n = number of units comprising the sample; c = number of sample units giving values over m or between m and M; m = minimum level acceptable; M = maximum level tolerable

Food Safety and Standards (Food Products Standards and Food Additives) Third Amendment Regulations 2017 gives the microbiological requirements for various fishery products for *Escherichia coli*. The method prescribed is IS: 5887 Part 1 or ISO 16649-2.

Food Category	Sampling plan		Limits (cfu/g)	
	n	c	m	M
Chilled/frozen finfish, and crustaceans	5	3	11	500
Chilled/ frozen cephalopods	5	3	11	500
Chilled /frozen cephalopods	5	0	20	
Chilled /frozen bivalves	5	0	46	
Live bivalve molluscs	5	0	230/100g	700/100g
Frozen cooked crustaceans/ frozen heat shucked molluscs	5	2	1	10

[Table Contd.

Contd. Table]

Food Category	Sampling plan		Limits (cfu/g)	
	n	c	m	M
Dried/salted and dried, and smoked fishery products	5	0	20	
Fermented fishery products	5	2	4	40
Smoked fishery products	5	3	11	500
Accelerated freeze dried fishery products	5	0	20	
Fish mince / surimi and analogues	5	0	20	
Fish pickle	5	0	20	
Battered and breaded fishery products	5	2	11	500
Convenience fishery products	5	2	1	10

Where n = number of units comprising the sample; c = number of sample units giving values over m or between m and M; m = minimum level acceptable; M = maximum level tolerable

9.4.1.1. MPN method

Reagents

1. Butterfield buffered peptone water
2. Lauryl sulphate tryptose (LST) broth
3. Brilliant green lactose bile (BGLB) broth, 2%
4. EC broth
5. Levine's eosin-methylene blue (L-EMB) agar
6. Plate count agar (PCA)
7. Violet red bile agar (VRBA)
8. 4 methyl-umbelliferyl-b-D-glucuronide

Presumptive test for coliform bacteria

Aseptically weigh 50 g of fish sample and homogenize in 450 ml Butterfield phosphate-buffered diluent for 2 min. Prepare decimal dilutions with 90 ml of the same diluent. Transfer 1 ml portions from each dilution to 3 LST tubes for 3 consecutive dilutions. Incubate the tubes for 48 h at 35°C, and examine the tubes at 24 ± 2 h for acid and gas production inside the Durhams tube. Re-incubate the negative tubes for an additional 24 h.

Confirmatory test for coliforms

Transfer a loopful of suspension from LST tube to the BGLB broth and incubate for 48 h at 35°C. Examine the gas production and record it. Calculate the most probable number (MPN) of coliforms based on proportion of confirmed LST tubes for 3 consecutive dilutions.

Confirmatory test for fecal coliforms

Gently agitate each LST tube and transfer a loopful of each suspension from the LST tubes simultaneously to the EC broth. Incubate the EC tubes for 48 h at 44.5 ± 0.2°C in a water bath. Examine the gas production at 24 h. If negative, re-examine it again at 48 h. Calculate the fecal coliform MPN based on the results of this test.

Confirmatory test for *E. coli*

Streak a loopful of suspension from each positive EC broth tube to L-EMB agar and incubate the tubes for 18-24 h at 35°C. Examine the plates for typical *E. coli* colonies, which appear as dark centered and flat, with or without metallic sheen. Transfer two typical colonies from each L-EMB plate to PCA slants for morphological and biochemical tests. Incubate the PCA slants for 18-24 h at 35°C.

9.4.1.2. Direct plating method for total coliforms

Prepare the Violet red bile agar (VRBA) and pasteurize by boiling for 2 min. Approximately homogenize 25 g of fish sample for 1 min in 225 ml Butterfield phosphate-buffered dilution water. Prepare serial tenfold dilution in butterfield diluent. From that, transfer 1 ml aliquots of each dilution into Petriplates. Then, pour 10 ml VRBA tempered to 48°C into plates. Swirl the plates to mix, and allow to solidify. Overlay the agar with additional 5 ml VRBA to prevent surface growth and spreading of colonies, and allow to solidify. Use 100 mg of 4 methyl-umbelliferyl-b-D-glucuronide (MUG) per ml in the VRBA overlay to find *E. coli* among coliforms,and observe for fluorescent colonies under longwave UV light. Add aliquot of up to 4 ml of dilution, when deeper plates are used and then add 15 ml of VRBA. Count the purple-red colonies having 0.5 mm or larger diameter and surrounded by zone of precipitated bile acids. Determine the number of coliforms/ g by multiplying percentage of tubes confirmed as positive by original VRBA count. For confirmation, select the colonies and transfer to each tube of BGLB broth. Incubate the tubes at 35°C for 24 and 48 h to examine for gas production. Colonies producing gas are confirmed as coliforms.

Calculation

$$\text{Total coliforms (CFU/g)} = \frac{\text{No. of colonies} \times \text{dilution} \times 10}{\text{Wt. of the sample, g}}$$

9.4.1.3. Direct plating method for *E. coli*

Reagents

1. Tergitol 7 Agar
2. 2,3,5- Triphenyl Tetrazolium Chloride (TTC), 1%
3. Physiological saline, 0.85%

Procedure

Aseptically weigh 25g of fish sample and homogenize with 225 ml of 0.85% saline to obtain 10^{-1} dilution. Prepare serial decimal dilutions using 9 ml of diluent. Spread plate 0.1 ml of sample onto pre-poured Tergitol 7 agar plates. Invert the petriplates and incubate at 35°Cfor 24-48h. The colonies of *E. coli* appear as yellow. Count the colonies between 25 – 250 range, calculate and express as CFU per gram.

Calculation

$$E.\ coli\ (\text{CFU/g}) = \frac{\text{No. of colonies} \times \text{dilution} \times 10}{\text{Wt. of the sample, g}}$$

Biochemical confirmation of *E. coli*

Test	Reaction	Test	Reaction
Nitrate reduction	+	Citrate	-
Cytochrome oxidase	-	Voges-Proskauer	-
Gram-negative, short rod	+	KCN	-
Fermentative (TSI)	+	Indole	+
Mannitol	+	Acetate	+
Lactose	+	Adonitol	-
Malonate	-	Cellobiose	-
H_2S	-	Urease	-
Glucose, gas	+	ONPG test	+
Arabinose, acid	+	Lysine decarboxylase	80% +
Methyl red	+		

A multiplex PCR assay is developed for the detection of *E. coli* using selected primers designated on the species- and strain-specific genes such as *hlyA, LT1*, and *phoA*. The internal regions amplified have a product size of 165, 275, and 903 bp, respectively. This MPCR assay has a sensitivity of detecting four cells of *E. coli* within 10h in artificially contaminated shrimp tissue homogenate. It can detect different strains of *E. coli* (EHEC, EPEC, ETEC, and EAEC) from different samples (Jeyasekaran *et al.*, 2014).

9.4.2. Enumeration of *Staphylococcus aureus*

S. aureus are gram-positive, non-motile, non-sporing, facultative anaerobic, catalase-positive cocci belonging to the family Micrococcaceae. They are ubiquitous, but they are commonly present in human beings and animals. About 10 to 30% of healthy humans carry these organisms in the throat and nose and hence referred to as indicators of personnel hygiene. Coagulase-positive *S. aureus* is of more importance, as they are highly pathogenic to humans and produce a toxin called enterotoxin, which is stable even at 100°C. It is highly vulnerable to destruction by heat treatment and nearly all sanitizing agents. Thus, the presence of this bacterium or its enterotoxins in processed foods is an indication of poor sanitation. Hazards from *S. aureus* are controlled by minimizing time/temperature abuse of aquatic food, especially after cooking and by following proper hygienic practices by food handling. Direct plate count method is suitable for the enumeration of *S. aureus*. The maximum allowable number of *S. aureus* in aquatic food products is 100 /g. Baird Parker agar medium is a highly selective medium used for the isolation of *S. aureus* from foods. The selectivity of this medium is due to the addition of sodium pyruvate and potassium tellurite. *S. aureus* produces two diagnostic characteristics on the opaque medium. The first one is that they produce clear zones by lipolysis or proteolysis. The second one is that zones appear within the clear zones, probably caused by lipase or lecithinase. *S. aureus* forms black colonies on this medium, as they reduce potassium tellurite to telluride.

Food Category	Sampling plan		Limits		Analytical reference method	Stage where the criterion applies
	n	c	m	M		
Coagulase-positive staphylococci	5	2	100 cfu	1000 cfu	EN/ISO 6888-1 or 2	End of the production process

Where n = number of units comprising the sample; c = number of sample units giving values over m or between m and M; m = minimum level acceptable; M = maximum level tolerable

The microbiological standards for fishery products set out in Commission Regulation (EC) No 2073/2005 of 15 November 2005 gives the microbiological criteria for coagulase positive Staphylococci in shelled and shucked products of cooked crustaceans and molluscan shellfish.

Food Safety and Standards (Food Products Standards and Food Additives) Third Amendment Regulations 2017 gives the microbiological requirements for various fishery products for coagulase positive Staphylococci The method prescribed is IS: 5887 Part 2 or IS 5887 Part 8 (Sec 1)/ ISO : 6888-1 or IS: 5998 Part (Sec 2)/ ISO 6888 – 2.

Food Category	Sampling plan		Limits		Stage where criterion
	n	c	m	M	applies
Frozen cooked crustaceans/ frozen heat shucked molluscs	5	2	1×10^2	1×10^3	End of manufacturing process
Fermented fishery products	5	1	1×10^2	1×10^3	End of manufacturing process
Smoked fishery products	5	2	1×10^2	1×10^3	End of manufacturing process
Accelerated freeze dried fishery products	5	0	10^2		End of manufacturing process
Fish mince / surimi and analogues	5	2	1×10^2	1×10^3	End of manufacturing process
Fish pickle	5	1	1×10^2	1×10^3	End of manufacturing process
Battered and breaded fishery products	5	1	1×10^2	1×10^3	End of manufacturing process
Convenience fishery products	5	2	1×10^2	1×10^3	End of manufacturing process
Powdered fish based products	5	2	1×10^2	1×10^3	End of manufacturing process

Where n = number of units comprising the sample; c = number of sample units giving values over m or between m and M; m = minimum level acceptable; M = maximum level tolerable

9.4.2.1. Direct plating method

Reagents

1. Phosphate buffered saline (PBS)
2. Baird-Parker (BP) medium

3. Egg yolk emulsion
4. Trypticase (tryptic) soy agar (TSA)

Procedure

Aseptically weigh 25 g of fish sample, blend for 2 min in a homogenizer and transfer to 225 ml of phosphate buffered saline (PBS). Prepare serial dilutions with the same diluent. From each dilution, transfer aseptically 0.1 ml of the sample suspension into Baird-Parker agar plates. Spread the inoculum over the surface of agar plate, using sterile bent glass streaking rod. Retain the plates in upright position until inoculum is absorbed by agar. Invert the plates and incubate for 48 h at 35°C. The colonies of *Staphylococcus aureus* appear as black, convex, 1.0 to 1.5 mm diameter, narrow white entire margin and surrounded by a mark of clearing 2 to 5 mm in width. The colonies may produce wide, opaque zones extending into the cleared medium on longer incubation. Coagulase-negative Staphylococci occasionally grow but rarely produce clearing. This opaque zone is an indication of the presence of coagulase. Count the positive colonies, calculate and express the results as number of CFU per gram. Pick up three or five typical colonies from BP agar plates and sub-culture onto trypticase soy agar (TSB) plates to purify the colonies for further biochemical identification.

Calculation

$$\text{Total staphylococci (CFU/g)} = \frac{\text{No. of colonies} \times \text{dilution} \times 10}{\text{Wt. of the sample, g}}$$

Biochemical confirmation of *Staphylococcus aureus*

Characteristics	S. aureus	S. epidermidis	Micrococci
Catalase activity	+	+	+
Coagulase production	+	−	−
Thermonuclease production	+	−	−
Lysostaphin sensitivity	+	+	−
Anaerobic utilization of glucose	+	+	−
Anaerobic utilization of mannitol	+	−	−

[a]+, Most (90% or more) strains are positive; -, most (90% or more) strains are negative

A multiplex PCR assay is developed for the simultaneous detection of *Staphylococcus aureus* using selected primers designed to target genus-specific *gap* gene, species specific *nuc* gene and enterotoxin-producing *EntC1* gene at product sizes 933 bp, 273 bp and 531 bp, respectively. The assay has a sensitivity to detect 10 cells of *S. aureus* and 100 pg of genomic DNA of *S. aureus* (Jeyasekaran *et al.*, 2011). The primers and product sizes of the target genes are given below:

Target genes	Primers	Product size
gap	F- ATG GTT TTG GTA GAA TTG GTC GTT TA R- GAC ATT TCG TTA TCA TAC CAA GCT G	933
nuc	F- GCG ATT GAT GGT GAT ACG GTT R- AGC CAA GCC TTG ACG AAC TAA AGC	273
EntC1	F- ACA CCC AAC GTA TTA GCA GAG AGC C R- CCT GGT GCA GGC ATC ATA TCA TAC C	531

9.4.3. Enumeration of Fecal Streptococci

Streptococci are gram-positive, non-motile, non-sporing, facultatively anaerobic and catalase-negative cocci. They are important human pathogens. They grow only in media containing fermentable carbohydrates or those enriched with blood or serum. The primary habitats of fecal streptococci are the intestinal tract of the humans, animals, and birds. Their presence is an indication of fecal contamination like *E. coli* in foods. Since they are resistant to lower and higher temperature processing treatments, they are the best indicators of fecal contamination in processed foods especially frozen foods. Important sources of Streptococci contamination are polluted waters and unclean workers.

9.4.3.1. Direct plating method

Reagents

1. Kenner Fecal (K.F) Agar
2. 2,3,5- Triphenyl Tetrazolium Chloride (TTC), 1%
3. Physiological saline, 0.85%

Procedure

Aseptically weigh 25g of the sample and homogenize with 225 ml of 0.85% saline to obtain 10^{-1} dilution. Prepare serial decimal dilutions using 9 ml of diluent. From that, spread plate 0.1 ml of sample onto pre-poured KF agar plates. Invert the petriplates and incubate at 35°Cfor 24-48h. The colonies of fecal streptococci appear as red pinpoint colonies. Count the colonies between 25 – 250 range, calculate and express as CFU per gram.

Calculation

$$\text{Fecal streptococci (CFU/g)} = \frac{\text{No. of colonies} \times \text{dilution} \times 10}{\text{Wt. of the sample, g}}$$

Biochemical confirmation of fecal streptococci

Tests	Reactions	Result
Catalase	No effectiveness	Negative
V.P	No change in colour	Negative
Gelatinase	Not liquefied	Negative
Blood agar (hemolysis)	Clear zone	Positive
Nitrate	Not reduced	Negative
Sorbitol	Fermentative	Positive
Trehalose	Fermentative	Positive
Lactose	Fermentative	Positive
Maltose	Fermentative	Positive
Dextrin	Fermentative	Positive
Mannitol	Fermentative	Positive

9.5. SPECIFIC ORGANISMS ASSOCIATED WITH PROCESSED FISHERY PRODUCTS

Microbiological assessment of processed fishery products differs with the nature of the product, the presence of specific group of microorganisms, and their survival in the relevant storage condition. Some specific organisms associated with the canned, salted, and MAP/fermented aquatic food products are discussed in brief.

9.5.1. Canned fishery products

The spoilage incidence in canned foods is limited. The spoilage in cans progresses from normal to flipper, springer, soft swell, and hard swell. Microbial spoilage and hydrogen produced by the interaction of acids in food products with the metal of the can are the principal causes of swelling. Some microorganisms do not produce gas and cause no abnormal appearance of the can, but cause spoilage of the product. Microbial spoilage is also due to leakage or under processing. Microbial spoilage assessment is by sterility testing in canned aquatic food products.

9.5.1.1. Sterility Testing

Eight-tube techniques commonly employed for the microbial examination of canned foods give a good idea about the type of organisms responsible for spoilage. Generally, a large number of cans are tested bacteriologically to obtain reliable results. When the cause of spoilage is clear, testing 4-6 cans may be adequate. The presence of viable but dormant organisms in unspoiled cans is examined bacteriologically, in which case, the procedure is the same except that the number of cans must be more.

Reagents

1. Glucose tryptone broth
2. PE-2 broth or thioglycolate broth
3. Crystal violet or gram stain
4. 4% iodine in 70% alcohol

Can preparation

Remove the labels and mark the side of the cans suitably with a marking pen. Separate the cans to be tested by code numbers and record the size of the container, code, product, condition, evidence of leakage, pinholes or rusting, dents, buckling, or other abnormality. Classify the cans according to the degree of spoilage signs like flipper, springer, soft swell, and hard swell.

Examination of can contents

Analyze the springer and swells immediately. Place the normal cans including flipper in the incubator at 37°C and examine at frequent intervals for 14 days. When the cans become increasingly swollen, make note of it and when swelling no longer progresses or when can becomes a hard swell, examine the contents. When thermophilic spoilage suspected or when the cans are to be held at high temperature in storage or at transit, incubate the cans at 55°C for 7 days.

Opening of can

Open the can in a sterile environment. Keep the hard swell cans in the refrigerator before opening. To sanitize, use 4% iodine in 70% alcohol and then wipe off with a towel (do not flame). For cans other than those with hard swell, scrub the entire uncoded end of the can with a brush using warm water and soap, rinse, and dry.

Flame the end thoroughly or flood with the iodine-alcohol solution and burn it off. Sterilize the can opener by flaming. While puncturing swollen can, hold the mouth of a sterile test tube using forceps to capture the escaping gas. Flip the mouth of the tube to the flame of the Bunsen burner. A slight explosion indicates the presence of hydrogen. Immediately, turn the tube upright and pour in a small amount of lime water. A white precipitate indicates CO_2. Make an opening large enough to permit removal of the sample.

Microbial examination

Remove large portions using wide-mouthed pipettes or spatula in case of solid pieces, from the center of the can. Transfer 1-2 ml or g of product into four tubes of PE-2 medium or thioglycolate medium, and four tubes of glucose tryptone broth, and incubate them as given below:

Medium	No. of tubes	Temperature (°C)	Time of Incubation (days)
PE-2/ TG	2	37°C	4 – 5
PE-2/ TG	2	55°C	1 – 3
GTB	2	37°C	4 – 5
GTB	2	55°C	1 – 3

Note:

a. Overlay the PE-2/ TG medium tubes with sterile liquid paraffin

b. Pasteurize one complete set of tubes at 100°C for 10-15 min to get an idea as to whether the organisms are spore formers or not.

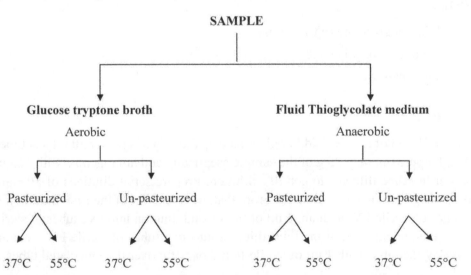

This technique leads to the following results

1. Whether the organism is aerobic or anaerobic
2. Whether the organism is spore former or non-spore former
3. Whether the organism is mesophilic or thermophilic
4. Whether the organism is gas former or non-gas former

Microscopic examination

Prepare a direct smear from the contents, air dry, heat fix, and stain with methylene blue or crystal violet, and examine under the microscope. Record the types of bacteria seen and calculate the total number per field.

9.5.1.2. Enumeration of Anaerobic bacteria

Anaerobic microorganisms like *Clostridium* sp. may be present in the canned fishery products. They cause major quality problems in vacuum packaged fish and smoked fish. *C. botulinum* produces a heat-labile toxin known as botulinal toxin, which is neurotoxic. The food poisoning caused by *C. botulinum* is known as botulism. *C. botulinum* Type E can grow and produce toxin of at low temperature 4°C. These organisms require an anaerobic environment for their growth. Anaerobes are cultured by different methods with the addition of reducing agents in the culture medium, the addition of alkaline pyrogallol, the use of candle jar, the use of MacIntoch and Filde's jar, and the use of the gas pack.

Reagents

1. Fluid thioglycolate (FY) medium
2. Physiological saline, 0.85%
3. Liquid paraffin

Procedure

Prepare 10 ml of sterile fluid thioglycollate medium in a series of 30 ml test tubes. Weigh approximately 25 g of the sample aseptically and homogenize with 225 ml of sterile saline diluents to get 10^{-1} dilution. Prepare serial dilutions of 10^{-2} and 10^{-3} using 9 ml of saline diluents. From that, transfer 1 ml of the first dilution into 5 tubes of sterile FY medium, 1 ml of the second dilution into five tubes of sterile FY medium and 1 ml of the third dilution into five tubes of sterile FY medium. Overlay the tubes with 1 ml of sterile liquid paraffin over the inoculated tubes in order to prevent the entry of oxygen from the atmosphere. Incubate the tubes at 35°C for 24-48 h. Examine the growth (turbidity), which is an indication of the presence of anaerobic bacteria. The disappearance of pink color in the surface layer of the medium due to the reduction reaction is a sign of anaerobic bacteria. Report the number of anaerobic bacteria present in the sample using a 5-tube MPN Table.

9.5.1.3. Enumeration of anaerobic sulfite reducing clostridia

Anaerobic sulfite reducers produce H_2S in aquatic food, particularly cuttlefishes, and squids resulting in sulfite spoilage. The most probable number (MPN) using the 3-tube technique is employed most commonly for the enumeration of anaerobic sulfite reducing clostridia.

Reagents

1. Differential Reinforced Clostridial Medium (DRCM)
2. Physiological saline, 0.85%

9.5.1.3.1. MPN method

Weigh approximately 10 g of the sample and dilute with 90 ml of sterile normal saline. Prepare serial dilutions with the same diluent. From that, inoculate 1ml of the first dilution into three tubes of DRCM, 1 ml of the second dilution into another three tubes of DRCM, and 1 ml each of the third dilution to the next three tubes of DRCM. Incubate the tubes at 37°C in a serological water bath up to 4 days. Observe the tubes for blackening after 48 h. Note down the number of positive tubes down and determine the number of sulphite reducing clostridia in the sample using a 3-tube MPN table.

9.5.1.3.2. Plating method

Weigh approximately 10 g of the sample and homogenize with 90 ml of sterile physiological saline, to get 10^{-1} dilution. Make serial dilutions using 9 ml of the same diluent. From the first three dilutions, transfer 1 ml into each sterile petriplate. Cool the sterile differential reinforced clostridial agar to 45°C, pour, and allow the agar to set. Overlay 15 ml of the same media and allow the agar to set. Incubate the plates in an anaerobic jar in an inverted position for 48 h at 35°C. Observe the plates for black colonies and count as sulfite reducing clostridia.

Calculation

No. of sulphite reducing clostridia / g = Average count x dilution factor

9.5.2. Salted fishery products

9.5.2.1. Enumeration of Halophilic bacteria

Halophiles are the organisms that require high salt concentrations for their growth, and generally present in salt-cured fishery products. Halophiles belong to three groups based on the salt requirement. Slightly halophiles require 5% salt, moderately halophiles require 5-20% salt, and extremely halophiles require 20-40% salt for their growth.

Reagents

1. 3% saline diluent for slightly halophiles
2. 10% salinediluent for moderately halophiles
3. 25% saline diluent for extremely halophiles
4. Plate count agar with 3%, 10%, and 25% salt

Procedure

Weigh approximately 25 g of the sample and homogenize with 225 ml of the respective diluent. Make the serial dilutions with 9 ml of the appropriate diluent. From that, transfer 0.1 ml from each dilution in the pre-poured plate count agar containing respective concentration of salts and do spread plating. Incubate the plates at 35°C for 24-48h. Count the colonies within the range of 25-250, and express as CFU per gram.

Calculation

$$\text{Halophiles (CFU/g)} = \frac{\text{No. of colonies} \times \text{dilution} \times 10}{\text{Wt. of the sample, g}}$$

9.5.2.2. Enumeration of Molds

Molds are present in the foods in which the environment is less favorable for bacterial growth. For instance, low pH, low moisture, high salt or sugar content, low storage temperature, presence of antibiotics, and exposure to irradiation.

Molds cause discoloration in food surfaces, development of off-odors and off-flavors, various degrees of spoilage, alter substrates allowing outgrowth of pathogenic bacteria, and produce mycotoxins in certain instances. Molds occur in salted, dried, and smoked fishery products commonly.

Food Safety and Standards (Food Products Standards and Food Additives) Third Amendment Regulations 2017 gives the microbiological requirements for various fishery products for yeast & mould The method prescribed is IS: 5403/ ISO 21527.

Food Category	Sampling plan		Limits		Stage where criterion
	n	c	m	M	applies
Dried / Salted and ied fishery products	5	2	100	500	End of manufacturing process
Fermented fishery products	5	0		100	End of manufacturing process
Fish pickle	5	0		100	End of manufacturing process
Battered and breaded fishery products	5	0		100	End of manufacturing process
Powdered fish based products	5	0		100	End of manufacturing process

Where n = number of units comprising the sample; c = number of sample units giving values over m or between m and M; m = minimum level acceptable; M = maximum level tolerable

9.5.2.2.1. Plating method

Reagents

1. Yeast extract - Dextrose – Chloramphenicol agar (YDCA) or Potato dextrose agar (PDA)
2. Physiological saline, 0.85%

Procedure

Weigh approximately 25 g of the sample and homogenize with 225 ml of physiological saline to get 10^{-1} dilution. Make serial dilutions using 9 ml of the same diluent. From that, transfer 0.1 ml of sample from each dilution to the sterile petriplates. Pour sterile YDCA or PDA maintained at 45°C. Incubate the plates at 25°C for 5 days. Count the colonies and report. The upper countable limit for molds is 10 to 150 colonies.

Calculation

$$\text{Total molds (CFU/g)} = \frac{\text{No. of colonies} \times \text{dilution factor}}{\text{Wt. of the sample, g}}$$

9.5.3. Fermented fishery products

9.5.3.1. Enumeration of lactic acid bacteria

Lactic acid bacteria (LABs) are gram-positive rods, cocci or coccobacilli, non-motile, and non-spore forming bacteria. They are catalase and cytochrome oxidase negative. They produce lactic acid as a product of fermentative metabolism of sugars and often associated with fermented foods and carbohydrate-rich foods. They occur in aquatic food products often implicated in food spoilage. They are also considered useful in food preservation and as probiotic.

Reagents

1. de Man Rogosa and Sharpe (MRS) agar
2. Physiological saline, 0.85%

Procedure

Weigh approximately 10 g of the sample and homogenize with 90 ml of physiological saline to get 10^{-1} dilution. Make serial dilutions with 9 ml of the same diluent. From that, transfer 1 ml of the first three dilutions into sterile petriplates. Cool the sterile MRS agar to 45°C, pour, and allow the agar to set. Incubate the plates in an inverted position in a CO_2 incubator at 5% CO_2 for 48 h at 35°C or in an anaerobic jar with 5% CO_2 for 48-72 h at 35°C or at room temperature (30°C) for 5-6 days. Observe the plates for white colonies of 2-3 mm diameter and count as lactic acid bacteria.

Calculation

$$\text{Total lactic acid bacteria (CFU/g)} = \frac{\text{No. of colonies} \times \text{dilution factor}}{\text{Wt. of the sample, g}}$$

CHAPTER 10

ANALYSIS OF
CHEMICAL CONTAMINANTS

This chapter gives a broad overview of the different chemical contaminants associated with aquatic food. The potential persistent environmental pollutants (PEPs) are organohalogen compounds, polycyclic aromatic hydrocarbons, and heavy metals.

10.1. ORGANOHALOGEN COMPOUNDS

Organohalogen compounds (OC's) are a group of stable carbon-based compounds having low volatility and high lipid solubility. OC's can enter the aquatic environment from industrial wastes, fish farm activities, agricultural cultivations, and untreated sewage discharge. OC's adsorb onto suspended particulates and deposit on the sediments. Aquatic organisms accumulate these compounds in their fatty tissues through direct contact with the water, suspended particles, and bottom sediments. Dietary intake contributes up to 90% of human exposure to OCs. Their presence in blood or mother's milk provides a reliable marker of human exposure to these compounds. The list of major OC's present in aquatic foods is given.

1. Polychlorinateddibenzo dioxins (PCDDs) - dioxins
2. Polychlorinateddibenzofurans (PCDFs) - dioxins
3. Polychlorinated biphenyls (PCBs)
4. Organochlorine pesticides (OCPs)

10.1.1. Dioxins

Polychlorinated dibenzo-p-dioxins (PCDDs) and polychlorinated dibenzofurans (PCDFs) are the two groups of persistent organic pollutants (POPs) generally referred to as 'Dioxins'. PCDFs and PCDDs formed as by-products of a wide variety of chemical industry and combustion processes that contain chlorine and chlorinated aromatic hydrocarbon sources. Waste incineration is the largest contributor to the release of PCDDs and PCDFs into the environment as a result of incomplete combustion.

Among the 210 possible congeners, seven 2,3,7,8-substituted PCDDs and ten PCDFs are considered the most persistent and toxic congeners. Their toxic properties similar to 2,3,7,8-tetrachloro dibenzo-p-dioxin (TCDD), which is the most toxic congener among these compounds. Minimum permissible level (MRL) for dioxins in fish and seafood is 3.5 pg of toxic equivalents (TEQs)/g fresh weight for PCDD/Fs and 6.5 pg of toxic equivalents (TEQs)/g fresh weight for the sum of PCDD/Fs and dioxin-like PCBs, except for eels and their products. MRL for dioxins in eels and their products is 10 pg of TEQs/g fresh weight (EC No. 1881/2006). The action level prescribed for dioxins and furans in farmed fish and fishery products is 1.5 pg of TEQS/g (2014/633/EU).

Analysis of these compounds in fatty seafood tissues requires three main steps: extraction, cleanup, and GC separation. GC-HRMS provides high specificity and sensitivity at concentration levels down to femtograms per gram, but it is a relatively expensive technique. Alternate techniques recently validated are GC-coupled to ion trap mass spectrometry and two dimensional GC coupled to micro electron capture detection and time of flight.

10.2. POLYCHLORINATED BIPHENYLS (PCBS)

Polychlorinated biphenyls (PCBs) are the environmental persistent pollutants, commonly referred to as synthetic organic compounds. They are present in electrical equipment as dielectric insulating media. They tend to accumulate in body fat in detectable levels in marine organisms, from mollusks to fish. Among the 209 PCB congeners, 12 are referred to as "dioxin-like" PCBs, as they exhibit toxicity similar to that of PCDDs/PCDFs. The maximum level prescribed for the sum of PCB 28, 52, 101, 138, 153 and 180 for the fish and fishery products is 75 ng/g wet weight, for the world caught freshwater fish is 125 ng/g we weight, for eel products is 300 ng/g wet weight (EC No. 1881/2006). The PCBs are determined using GC fitted with electron capture detection. The individual PCB congeners are determined using GC coupled with mass spectrometry. High-resolution mass spectrometry (HRMS) presents high specificity and sensitivity, and hence it is a reference method for the determination of trace levels of non-ortho and mono-ortho PCBs.

10.2.1. Gas Chromatographic Analysis

Reagents

1. Petroleum ether
2. Anhydrous sodium sulfate
3. Acetonitrile
4. Ethyl ether
5. Sodium chloride
6. Florisil Bond Elut column – SPE cartridges
7. PCB standard solution 7 (PCBs 28, 52, 101, 138, 153, 180 and 209)

Extraction

Weigh approximately 25-50 g of fish tissue, grind in a tissue blender and transfer to the conical flask. Add 200 g of sodium sulfate and mix well. Then, add 150 ml of petroleum ether and agitate for 10 min. Filter the content through the vacuum filtration flask. Re-extract the residues once again with 100 ml of petroleum ether and filter. Combine the extracts and allow them to pass through sodium sulfate taken over the filter paper in a funnel. Evaporate the petroleum ether to dryness using a rotary flash evaporator. Weigh the extracted lipid.

Cleanup

Take approximately < 3 g of the lipid for cleanup into a 125 ml separating funnel. Add 15ml of petroleum ether and 30 ml of acetonitrile saturated with petroleum ether. Shake the content for 1 min and allow the layers to separate. Drain the acetonitrile drained into a 1L separating funnel containing 650 ml of water, 40 ml of saturated sodium chloride solution, and 100 ml of petroleum ether. Extract the petroleum ether left out in the separating funnel thrice with 30 ml portions of acetonitrile saturated with petroleum ether and shake for 1 min each time. Combine all the extracts in another separating funnel and shake for 1 min. Allow the layers to separate and drain the aqueous layer into another 1 L separating flask. Add 100 ml of petroleum ether, shake for 15 sec, and allow the layers to separate. Discard the aqueous layer, combine the petroleum ether in the original 1L separating flask, and wash with 2 x 100 ml portions of distilled water. Drain the petroleum ether layer through sodium sulfate taken over the filter paper in a funnel. Evaporate the content using a rotary flash evaporator and re-constitute with little ether for transfer to Florisil bond Elutcolumn.

Florisil column cleanup

Place the Florisil Bond Elut column in the SPE system and pre-wet the column with 5 ml of petroleum ether. Transfer about 2 ml of sample extract to the column and elute with 2 x 5 ml portions of petroleum ether. Concentrate the eluate in a rotary flash evaporator under the stream of nitrogen and finally re-constitute with 1-2 ml of petroleum ether for injection into the GC.

GC analysis

Gas chromatograph equipped with electron capture detector.

Column: Capillary column SPB 608

Injector temperature: 300°C

Detector temperature:300°C

Carrier gas: Nitrogen – 15 psi

Sample injection: 1 µl

Temperature gradient program

Temp.	Rate	Hold
150°C	0	4 min
290°C	8°C/min	10 min

Simultaneously, inject the PCB standard to identify the compounds based on the retention times. Quantify the PCBs with the help of the software.

10.3. ORGANOCHLORINE PESTICIDES (OCPs)

The term 'pesticide' refers to the chemicals that have different physical and chemical properties. The use of pesticides had resulted in the presence of their residues in the aquatic environment. Once released into the environment, they transform into a range of different products due to their susceptibility to biotic and abiotic degradation. Pesticides are more mobile, persistent, and toxic to non-target organisms than their parent compounds. OCPs detected in aquatic food are o,p' and p,p' isomers of 1,1,1 trichloro 2,2 bis (p-chlorophenyl) ethane (DDT), 1,1-dichloro-2,2'-bis (p-chlorophenyl) ethylene (DDE) and (1,1-dichloro-2,2-bis (p-chlorophenyl) ethane (DDD), hexachlorobenzene (HCB), α-, β-, γ- isomers of hexachlorocyclohexane (HCHs), chlordane (CHLs; oxychlordane, trans- and cis-nonachlor, trans and cis-chlordane and heptachlor epoxide), drins (aldrin, dieldrin and endrin). The maximum permissible limits (MRLs) for few pesticides as per USFDA guidelines: DDT/ DDE/ DDD – 5.0 ppm; benzene hexachloride – 0.3 ppm; chlordane/ chlordecone – 0.3 ppm; hexachlor/ hexachlor epoxide – 0.3 ppm; aldrin/dieldrin - 0.3 ppm.

10.3.1. Gas chromatographic analysis

Pesticide residues are extracted from fish tissue using petroleum ether. The anhydrous sodium sulfate removes water from the tissue and helps to disintegrate the sample. Pesticide residues are then isolated from lipid by partition between petroleum ether and acetonitrile. Most of the lipid is retained in petroleum ether while the residues partition into acetonitrile. The subsequent step partitions back the residues in acetonitrile into petroleum ether. The added water reduces their solubility in acetonitrile. Separate the residues in solution from sample co-extractives on a column of Florisil adsorbent by subsequent elution, and by way of increasing polarity, to sequentially remove the residues from the column.

Florisil column cleanup

Fix the Florisil Bond Elut SPE column in the manifold and pre-wet the column with 5 ml of petroleum ether. Transfer 2 ml of sample extract to the column. Elute the column with 2 x 5 ml portions of petroleum ether initially. Then, elute the column with 20 ml of 6% ethyl ether/ petroleum ether (6:94) eluant and collect the fraction. Concentrate the fraction in a rotary flash evaporator under the stream of nitrogen. Reconstitute with 1-2 ml of petroleum ether for injection into the GC. GC conditions similar to PCBs estimation.

10.4. POLYAROMATIC HYDROCARBONS (PAHs)

Polyaromatic hydrocarbons (PAHs) are environmental pollutants easily absorbed by animals. PAHs are produced during the incomplete pyrolysis of wood used to produce smoke and hence, considered as carcinogenic contaminants of processed food, particularly smoked foods. Benzopyrene is the first known PAH to have potent toxicity and carcinogenicity. PAHs are lipophilic compounds that can easily cross the biological membranes and accumulate in fatty tissues. Their breakdown in the liver leads to poly-hydroxy-epoxy PAHs, which form covalent adducts with DNA and initiates a mutagenic process in the cells.

The U.S Environmental Protection Agency (US-EPA) has identified 16 PAHs, as the most frequently occurring ones in nature. Aquatic organisms such as algae, mollusks, and primitive invertebrates metabolize PAHs to a lesser extent and accumulate them in high concentrations. On the other hand, fish and higher invertebrates metabolize PAHs and accumulate little or no PAHs. In 1988, the maximum limit specified by European Regulation for benzopyrene in foodstuffs treated with smoke flavorings is 0.03µg/kg. Chromatographic techniques such as HPLC and GC are the two techniques applied for the detection of PAHs in aquatic food.

10.4.1. Gas chromatographic analysis

Reagents

1. Alumina or Florisil Bond Elut SPE column
2. n-hexane
3. PAH standards mixture

   Pyrene, benzo(b)fluoranthene, benzo(k)fluoranthene, acenaphthylene, indeno (1, 2, 3-cd) pyrene, benzo(a)pyrene, acenaphthylene, fluorene and benzo(a) anthracene

Soxhlet extraction

Weigh 5 g of dried sample, homogenize and transfer it into the extraction thimble. Place the thimble in the extraction chamber of the Soxhlet unit. Extract with 200 ml n-hexane the lipid for 8 h. Concentrate the lipid extract in a rotary evaporator to dryness at 60°C. Then, re-constitute the residue with 5 ml of n-hexane, before column clean up.

SPE column clean up

Use, either the SPE alumina or the Florisil bond elute SPE column, for cleanup. Fix the column in the manifold and pre-wet with 5 ml of n-hexane for conditioning. Transfer 2 ml of the extract onto the column. Elute the column with 2 x 5 ml of n-hexane, and collect the eluate. Evaporate the eluate in a rotary evaporator to dryness. Re dissolve the dry eluate in 1 ml of n-hexane for GC analysis.

GC analysis

System: Gas chromatograph equipped with flame ionization detector

Column: fused capillary column- DB-5MS, 5% phenyl methyl polysiloxane

Dimensions: 30 m long ×0.25 mm i.d, 0.25 μm film thickness

Injector temperature: 300°C

Detector temperature: 300°C

Carrier flow rate: 4 ml/min.

Injection volume: 4 μl

Split ratio: 50:1

Temperature gradient programming

Temp.	Rate	Hold
60°C	–	2 min
170°C	40°C/min	–
220°C	10°C/min	–
290°C	5°C/min	10 min

Identify the PAHs based on a comparison of the retention times with those in a standard solution and quantify based on the corresponding areas of the respective chromatograms.

10.5. HEAVY METALS

Heavy metals accumulate in marine organisms, water and sediments mainly include arsenic (As), cadmium (Cd), mercury (Hg), and lead (Pb). Metals enter the aquatic environment through atmospheric deposition, erosion of the geological matrix, or from anthropogenic sources such as industrial effluents and mining wastes. Hg and As levels are high in the muscle of mullet, while Pb and Cd levels are high in the kidney of carp. Lobster, abalone, and snapper accumulate high amounts of As, Cd, and Hg, respectively. Provisional tolerable weekly intakes (PTWIs) for Cd, Pb, and Hg were established JECTA as 7, 25, and 1.6 μg/kg body weight, respectively. Maximum permissible limits of heavy metals range from 0.05 to 1.0 ppm for Cd, 0.3 to 1.5 ppm for Pb, and 0.5 to 1.0 ppm for Hg, depending on the type of the organisms.

10.5.1. Spectrochemical methods

Metals present in solutions quantified by spectrochemical methods. Sample preparation is conveniently divided into classical and microwave. Classical methods involve wet or acid decomposition. This method uses various mineral acids such as sulfuric, nitric, perchloric, and oxidizing agents, hydrogen peroxide to effect dissolution of the sample in an open or closed system. Microwave digestion involves the use of 2450 MHz electromagnetic radiation to digest samples in a Teflon or quartz container. Atomic absorption spectroscopy (AAS) with flame and graphite furnace is the most commonly used for the determination of heavy metals. Inductively coupled plasma (ICP) analyzer coupled with atomic emission spectroscopy (AES) as well as mass spectrometry (MS) is the method for the determination of multi-metals at a time. Levels of heavy metal in aquatic food are expressed in μg/g (ppm) or ng/g (ppb).

Reagents

Standard solutions for heavy metals

Prepare a series of the standard metal solutions by appropriate dilution of the stock metal solutions with distilled water containing 1.5 ml concentrated HNO3/ L. Primary standard solutions are commercially available. The preparation of standards is given below:

1. *Cadmium*: Dissolve 0.1 g cadmium metal in 4 ml of HNO_3. Add 8 ml of HNO_3 and dilute to 1L with distilled water (1 ml = 100 μg Cd)

2. *Chromium:* Dissolve 0.1923 g CrO_3 in distilled water. Acidify with 10 ml of HNO_3 and dilute to 1 L with distilled water (1 ml = 1 μg Cr)

3. *Cobalt:* Dissolve 0.1 g Co metal in minimum amount of HNO_3: water (1:1). Add 10 ml of HCl:water (1:1) and dilute to 1 L with distilled water (1ml = 100 μg Co)

4. *Copper:* Dissolve 0.1 g of Cu metal in 2 ml of conc. HNO_3. Add 10 ml of conc HNO_3 and dilute to 1 L with distilled water (1 ml = 100 μg Cu)

5. *Iron:* Dissolve 0.1 g iron wire in a mixture of 10 ml HCl:water (1:1) and 3 ml of conc. HNO_3 and dilute to 1 L with distilled water (1ml = 100 μg Fe)

6. *Lead:* Dissolve 0.1598g lead nitrate Pb $(NO_3)_2$ in a minimum amount of HNO_3: water (1:1). Add 10 ml of conc HNO_3 and dilute to 1L with distilled water. (1ml = 100 μg Pb)

7. *Manganese:* Dissolve 0.1 g of Mn metal in 10 ml of conc. HCl and mix with conc. HNO_3. Dilute to 1L with distilled water (1ml = 100 μg Mn)

8. *Nickel:* Dissolve 0.1 g Ni metal in 10 ml of hot conc. HNO_3. Cool and dilute to 1L with distilled water (1 ml = 100μg Ni)

9. *Zinc:* Dissolve 0.1 g of Zn metal in 20 ml of HCl: water (1:1) and dilute to 1L with distilled water. (1 ml = 100 μg Zn)

Mineralization

Weigh 5g of dried sample and subject to dry mineralization at 450°C in a muffle furnace. Alternatively, weigh 5 g of wet tissue in a glass tube with Teflon ends and add inorganic acids at the ratio of HNO_3 and H_2SO_4 (3:2) or HNO_3, H_2SO_4 and $HClO_4$ (2:1:1), to wet mineralize at 150°C under pressure. Otherwise, weigh 10g of homogenized wet tissue in a digestion flask, and add 10-15 ml of concentrated HNO_3. Keep it overnight. Then, add 2 ml of $HClO_4$ and digest it in a microwave digestor, until the solution is clear. If slightly yellow, add a little 10% H_2O_2, to complete digestion. Closed digestion used to digest metals, particularly mercury.

10.5.2. Atomic absorption spectrometry analysis

Introduce the element into a flame where it gets dissociated from its chemical bond into an unexcited un-ionized ground state, as individual atoms. At this stage, the element is capable of absorbing radiation at discrete lines of narrow wavelength. When the radiation of specific wavelength passes through the flame, the amount of light absorbed is directly proportional to the concentration of the element. The source of radiation is from the respective hollow metal cathode lamp. Two main types of atomization sources are available, namely flame and graphite furnace. The graphite furnace improves the sensitivity of atomic absorption 1000 times more than the flame atomic absorption.

Element	Wavelength (nm)	Flame (gases)	Sensitivity (mg/L)	Conc. Range (mg/L)
Cadmium, Cd	228.8	Air-Acetylene	0.025	0.05-2.00
Copper, Cu	240.7	Air-Acetylene	0.2	0.5-10.0
Chromium, Cr	357.9	Air-Acetylene	0.1	0.2-10.0
Copper, Cu	324.7	Air-Acetylene	0.1	0.2-10.0
Iron, Fe	248.3	Air-Acetylene	0.12	0.3-10.0
Lead, Pb	283.3	Air-Acetylene	0.5	1-20
Nickel, Ni	232.0	Air-Acetylene	0.15	0.3-10.0
Tin, Sn	224.6	Air-Acetylene	4.0	10.-200
Zinc, Zn	213.9	Air-Acetylene	0.02	0.05-2.00
Beryllium, Be	234.9	Nitrous oxide - Acetylene	0.03	0.05-2.00

10.5.3. Inductively coupled plasma – Atomic emission spectroscopy (ICP-AES) analysis

This method uses inductively coupled plasma generators as an atomization source for optimal emission spectroscopy. It has a high analytical sensitivity of analyzing 70 elements at a time, with the detection limits in the ppb range. The calibration graph is rectilinear over five orders of magnitude concerning analyte concentration.

Extraction and concentration

Metals form complexes by chelation reactions, which are required to extract a metal ion from the aqueous to the organic phase. The chelating agent is a pyrrolidine derivative of dithiocarbamic acid, and it forms non-selective complexes at pH 3-7. The method involves chelation with a mixture of equal amounts of ammonium dithiocarbamate between pH 4-5, which is adjusted by citrate buffer. The main disadvantage of metal carbamates in organic solvents is their poor stability. This problem can be resolved by evaporating the organic phase and dissolving the residue in the aqueous phase.

Transfer 100 ml of sample to a separating funnel and adjust the pH to 4.5 by the addition of 1 ml of citrate buffer. Add 1 ml of chelating reagent followed by 20 ml of methyl isobutyl ketone. Continue extraction by the addition of another 10 ml of methyl isobutyl ketone to the funnel. Combine the two extracts and add 0.5 ml of HNO_3. Mix for 1 min and let the tubes to stand for 15 min to decompose the metal carbamate. Add 3.5 ml of pure water and shake for 2 min to ensure complete back extraction.

10.5.4. Mercury analysis by cold vapour AAS

Potential human exposure to mercury (Hg) receives more attention. Mercury exists in inorganic and organic forms in water. Once released into the environment, inorganic Hg gets converted to organic Hg (methyl mercury, MeHg). MeHg is the most toxic chemical form, stable and easily absorbed from the diet. It is the most common form in seafood and could make up more than 90% of the total Hg. MeHg concentrations in fish and shellfish are approximately 1,000 to 10,000 times greater than in other foods, including cereals, potatoes, vegetables, fruits, meats, poultry, eggs, and milk. Fishes being top of the aquatic food chain contain MeHg levels 1 to 10 million times greater than those found in the surrounding water. MeHg damages the brain and parts of the nervous system, particularly of an unborn baby. The maximum permissible limit for Hg in aquatic food is 0.5-1.0 ppm.

Sample containing Hg^{2+} ions on treatment with $SnCl_2$ release elemental Hg. Liberated Hg vapor is drawn into the absorption cell of mercury analyser which is irradiated by a low-pressure mercury-vapor lamp. Mercury vapor absorbs the radiation at 253.7 nm and causes a change in transmittance, which is proportional to the Hg present in the sample solution.

Wet oxidation carried out for the conversion of all mercury monovalent, divalent, organic into divalent mercury, followed by reduction of divalent Hg to metallic Hg.

$$\text{Step 1: Bound Hg} \xrightarrow{\text{digestion}} \text{Inorganic mercury } (Hg^{2+})$$

$$\text{Step 2: Reduction } Hg^{2+} \xrightarrow[\text{HCl}]{\text{SnCl}_2} \text{Metallic mercury } (Hg°)$$

Mercury Analyzer

Mercury analyzer is a sensitive instrument designed for the determination of mercury residues at ng level. It is a cold vapor atomic absorption spectrophotometer that works on the principle that mercury vapor (atoms) absorbs resonance radiation at 253.7nm. Mercury analyzer consists of a low mercury lamp, as a radiation source, a filter, a detector, and a vapor generation system. Carrier air bubbles through the vapor generation system take the elemental mercury from the solution and then pass through the absorption cell.

Reagents

1. **$KMnO_4$, 1%**
   Dissolve 5 gof $KMnO_4$ in distilled water and add 50 ml of H_2SO_4 and make up the volume to 500 ml with distilled water.

2. NaOH, 20%

3. SnCl, 20%

4. Dissolve 20 g of stannous chloride 10 ml of hot HCl and make up to 100 ml with distilled water

5. Sulfuric acid, 18 M

6. HCl, 10 M

7. HNO_3 acid, 16 M

8. **Diluting acid solution**

Mix 100 ml of 16 M HNO_3 and 50 ml of 18 M H_2SO_4 with 850 ml of distilled water

9. **Standard mercury solution**

Stock (100μg/ml): Dissolve 0.1354 g of $HgCl^{2+}$ in 1L of 1 N HCl.

Working I (10μg/ml): Dilute 1 ml of the stock solution to 100 ml with 1N HCl

Working II (0.1μg/ml): Dilute 1 ml of the working I solution to 100 ml with 1 N HCl.

Operation

Adjust all control knobs on the front panel to the middle positions:

o   Zero: 5 turns from the clockwise extreme.

o   100% Coarse: Half-a-turn from the clockwise extreme.

o   Fine: Half-a-turn from the clockwise extreme.

o   Keep the Hold-Normal switch at 'Normal'

o   Keep the % T/Abs switch in %T position.

Push the Filter rod to the 'Filter' position. Switch "ON" the mains and allow for 1 min warm up. Put "ON" the lamp switch. Push the filter rod fully to the "filter-in" position and adjust "zero" with the ZERO control. Leave the filter rod in 'FILTER' position. Allow a warm-up time of about 30 min. Pull out the filter rod to the "OPEN" position and adjust "100% T" by rotating the coarse and fine control knobs.

Note:

i.   Never fill the traps with more liquid than specified, lest the liquid will get carried over to the next trap or the absorption cell.

ii.   Start the pump and allow the air to flow through for a minute. In the normal mode of operation, the digital meter reading should not be more than 0.5% T.

iii. Replace the traps at least once in three days.

iv.   Clean acid and moisture traps, and refill them with respective reagents daily.

Setting up of vapour generation system

The vapor generation system consists of a magnetic stirrer, a reaction vessel with a stirring paddle and four traps all housed in a wooden stand. The stand has two compartments. Magnetic stirrer and reaction vessel placed on the left-hand side and four traps on the right-hand side. The two bigger mounting holes are for two traps, filled with 20 ml of 1% $KMnO_4$ in 10% H_2SO_4. The smaller mounting holes are for acid trap filled with 4 ml of H_2SO_4 (1:1) and alkali trap filled with 4 ml of 20% NaOH. Vapour generation system and the electronic console are connected as per the manufacturer's instruction.

Procedure

Weigh 2 g of homogenized tissue into a conical flask and add 5 ml of 16 M HNO_3, 2.5 ml of 18 M H_2SO_4 and 1 ml of 10 M HCl. Cover the flask with glass marble, and the initial reaction takes place. After 15 min of reaction, place the flask in a boiling water bath for 40 min. Remove the flask from the water bath, cool, and makeup to 50 ml with distilled water. Pour 20 ml of digested sample into the reaction vessel, and add 3 ml of $SnCl_2$ solution and stir the solution and allow them to react for 5 min. Pass the generated mercury vapor through an absorption cell and detect at 253.7 nm by an atomic absorption spectrophotometer. Pipette out 0.2, 0.4, 0.6, 0.8 and 1.0 ml of standard mercury working II solutions (0.1 µg/ ml) containing 20, 40, 60, 80 and 100 ng of mercury into a series of test tubes and dilute to 20 ml with diluting acid solution and measure their absorbance in the same way, as in sample. Plot a standard mercury calibration graph with a concentration of mercury on the X-axis and absorbance in the Y-axis. Calculate the amount of mercury present in the given sample.

Calculation

$$\text{Mercury (ng/g)} = \frac{50 \times C}{W \times 20}$$

where, C = ng of mercury content; W – Wt. of the fish, g; 20 – aliquot taken, ml; 50 – made up volume

10.6. VETERINARY DRUGS

Antibiotics are substances produced by fungi or bacteria that are capable to kill microorganisms. Semi synthetic or synthetic substances lethal to microorganisms are termed as "antibiotics". Agents that cannot be considered under antibiotics include antihelmintica, antimycotica and hormones. Hence, all these substances are grouped under "Veterinary drugs". Fishes and shrimps are susceptible to a number of diseases by bacteria, virus or fungi. They are also prey to many parasites like protozoans, trematodes and helminths. Aquaculture bacterial and parasitic infections are often treated with veterinary drugs.

Few veterinary drugs are allowed for the use in aquaculture. Withdrawal times (ie. the time between the end of a treatment and the clearance of the drug from the organism) are an important issue. Continuous use of antibiotics has led to the emergence of resistance among bacteria, which forced the farmers to increase the dosage or shift to another drug. Some drugs are thus prohibited by many countries e.g. chloramphenicol and nitrofurans.

Some veterinary drugs have extremely long depletion times. Malachite green and its metabolite, leuco malachite green, deplete within 100 days from 700 to 15µg/kg in eel. Low levels of antibiotics are observed in commercial feed. Crabs are contaminated during processing by the commercial hand cream containing chloramphenicol. In general, veterinary drugs undergoes fast and extensive metabolism in the animal. Some drugs as indicators of the parent substance e.g., semicarbazide is the indicator metabolite of nitrofurazone.

The veterinary drugs for which MRLs not established include chloramphenicol, chloroform, chlorpromazine, colchicines, dapsone, dimetridazole, metronidazole, nitrofurans (including furazolidone) and ronidazole (EU No 37/2010). List of antibiotics and other pharmacologically active substances banned for using in shrimp aquaculture are chloramphenicol, nitrofurans (furaltadone, furazolidone, furylfuramide, nifuratel, nifuroxime, nifurprazine, nitrofurantoin, nitrofurazone), neomycin, nalidixic acid, sulphamethoxazole, and preparations thereof, chloroform, chlorpromazine, colchicines, dapsone, dimetridazole, metronidazole, ronidazole, ipronidazole, other nitroimidazoles, clenbuterol, diethylstilbestrol, sulfonamide drugs, fluroquinolones and glycopeptides. The maximum permissible limit for tetracycline is 0.1 ppm, oxytetracycline is 0.1 ppm, trimethoprim is 0.05 ppm, and oxolinic acid is 0.3 ppm (CAA, 2005).

a. Biological test system

The inhibition of the growth of microorganism by the presence of an antibacterial drug is utilized for their detection. These tests are not very selective. Good sensitivity is reported for penicillin. Nitrofurans or sulphonamides produce weak or even no inhibition. ELISA permits fast and highly selective screening and semi quantification of various veterinary drugs. This method is well suited for large number of processed samples. High selectivity is due to the recognition of an epitope of the veterinary drug. ELISA is still a 'one drug one test' approach.

b. Instrumental method

Liquid chromatography is more suited for the analysis of veterinary drugs. GC coupled with MS is still useful for hormone analysis. Samples are extracted and cleaned up prior to separation and quantification. Based on the group of veterinary drug, polar or non polar, highly buffered or strongly acid solvents are used for the extraction. There are some alternatives for liquid extraction: matrix solid phase dispersion extraction (MSPD), and accelerated solvent extraction (ASE). MSPD is laborious, while ASE is limited to thermostable analysis. Modern methods of veterinary drug clean up is solid phase extraction (SPE) e.g. C-18 cartridges. RP-SPE removes proteins and peptides which otherwise precipitate on the analytical column, reduce separation performance and cause signal suppression in MS. Anion or cation exchange cartridges are highly selective producing clean extracts, but not suitable for some veterinary drugs. Hence, less selective RP-SPE cartridges are preferred.

LC and UV detection are used for the assay of few drugs, but provide insufficient selectively and sensitivity. LC with fluorescence detector permit high selectivity and sensitivity but only a few numbers of veterinary drugs possess an inherent fluorescence e.g chinolones. Most of these drugs require pre or post column derivatization before detection and hence, not satisfactory. LC coupled with MS opened new frontiers for the analysis of veterinary drug residues. Interfaces like electrospray (ESI), atmospheric pressure chemical ionization (APCI) and photo ionization (PI) are used. Most of the veterinary drug analysis is done by LC coupled to tandem mass spectrometry (LC-MS-MS). Mass selection of a precursor ion, induced fragmentation and selective monitoring of one or more derived product ions gives good selectivity and sensitivity.

10.6.1. Chloramphenicol analysis by LC-MS/MS

This method is used for screening of chloramphenicol (CAP) residues in shrimp and shrimp products.

Materials & Reagents

1. LC-MS-MS system: UPLC-Waters Acquity™ Binary Gradient (Waters) with MS/MS, Qtrap (AB Sciex) having injection loop10 µl and syringe 100 µl in the autosampler. LC-MS-MS vials have 1ml dead volume.

2. UPLC column: Phenomenex 50x2mm 2.5µm

3. Methanol (MeOH) and MilliQ water, 1:1 v/v

4. Hexane and Carbon tetrachloride, 1:1 v/v

Preparation of NP standard solutions

1. Stock CAP solution (1000ppm): Accurately weigh 10 mg ± 0.02mg of CAP in a 10ml volumetric flask and add MeOH to the mark.

2. Intermediate solution of CAP-I (1ppm): Transfer 10 µl of stock CAP and make upto 10ml with methanol.

3. Intermediate solution of CAP-II (100ppb): Transfer 1000µl of intermediate stock I in a 10ml volumetric flask and add MeOH : MilliQ water (1:1, v/v) to mark. From this, prepare 0.1, 0.2, 0.3, 0.4 and 0.5 ppb (S-1 to 5) in Eppendorf tubes.

Preparation of the sample extract

Take approximately 100g of shrimp sample and homogenize at a low speed for 2 min in a blender. From that, weigh 2 ± 0.01 gm and add 0.03 ml of internal standard (D5 CAP – 20 ppb) and 2 ml of milli Q water and vortex for 1 – 2 min. Add 10 ml of ethyl acetate to the sample and vortex the sample for 10 min. Centrifuge the sample at 6000 x g at 4^{0}C for 10 min. Transfer the upper phase into a clean tube and evaporate the ethyl acetate solution to dryness under the stream of nitrogen at 40°C at 200 psi pressure for 20 min. Dissolve the dry residue in 1 ml of Hexane and Carbon Tetrachloride (1:1, v/v) and vortex for 10 sec. Add 1ml of Milli Q water and vortex for 10 sec. Transfer it into 2 ml Eppendorf tube and centrifuge for 5 min at 10000 rpm at 4°C. Transfer the upper phase into LC-MS/MS vial.

LC – MS- MS Analysis

Connect the UPLC, LC pump, Auto sampler, and Mass Spectrometer in series. Fix the injection loop on injection port in the auto sampler as 10 µl. Sample transfer tube in AS is 1ml vial. Standard tray contains 48 vials with septa and screw cap in two rows. Fix the UPLC column.

Set the auto sampler method as follows:

Needle height from bottom : 20 Mm

Injection volume and mode : 7.5 µl (Partial loop with needle overfill)

Post-injection valve switch time: 0.0min

Loop loading speed: 5 µl/s

Syringe speed : 8 µl/s

Flush/Wash source is wash bottle containing 50/50(v/v) methanol/water.

Flush/Wash volume : 250 µl

Flush speed: 250 µl/s

Tray temperature control: On. Temperature: 15°C

Column Oven Control: On. Temperature: 35°C

Set the method in UPLC as follows:

Mobile phase A: Acetonitrile

Mobile phase B: MilliQ Water

Allow mobile phase A/B at 50/50 (v/v) to run through UPLC system for at least 15 min at flow rate 0.3 ml/min to equilibrate the system. Then, set the gradient program as given below:

Time (min)	Flow rate /min	Mobile phase A%	Mobile phase B%
Initial	0.3	20	80
0.50	0.3	20	80
1	0.3	90	10
2	0.3	90	10
2.20	0.3	20	80
3	0.3	20	80

Retention times and molecular weight of the CAP		
Compound	Retention time (min)	MW
CAP	1.41	321

Set the TSQ Quantum method as follows:

Acquire time : 3 min

Ion Source : ESI

ESI capillary position : 90 degree

Calibration File Type : High Resolution

Ion Source

Ions spray voltage: 5500V

Ion source Gas 1(GS1) – 50

Ions source Gas 2 (GS 2) – 60

Lurtain Gas (CUR) – 10

Interface heater – ON

Polarity - Negative

Scan type: MRM

Set the MRM Conditions as follows:

Product ID	Parent (m/z) Q1 Mass	Product (m/z) Q3 Mass	Time (Ms)	Collision Energy (v)
CAP 1	321	152	200	-25
CAP 2	321	257	200	-22
CAP D5 (IS)	326	157	200	-20

Injection sequence

Set the injection volume as 10 μl

Inject a MeOH blank as the first and second injection

Inject a blank matrix as the third injection

Inject a positive control (Spiked sample with S-3 (0.3 ppb)) as the fourth injection

Inject S-1 (1 ppb) standard sample both to start and to end the bracket of samples.

Confirmation Criteria

Retention time of the analytes must agree with 5% of that of standard

Two ion ratios of the analytes must agree within 10% of that of standard.

10.6.2. Nitrofuran metabolites analysis by LC-MS/MS

This method is used for the screening of 3-amino-2-oxazolidinone (AOZ), 5-methyl-morpholino-3-amino-2-oxazolidinone (AMOZ), semicarbazide (SEM), and 1-aminohydantoin (AHD) residues, which are metabolites of furazolidone, furaltadone, nitrofurazone, and nitrofurantoin, respectively in shrimps.

Materials & Reagents

1. LC-MS-MS system: UPLC-Waters Acquity™ Binary Gradient (Waters) with MS/MS, Qtrap (AB Sciex) having injection loop 10 μl and syringe 100 μl in the autosampler. LC-MS-MS vials have 1ml dead volume.

2. UPLC column: ZORBAX Eclipse plus C18 4.6x150mm 5μm (Agilent)

3. 0.2 M HCl

4. 1 M Na_3HPO_4

5. 2 M NaOH

6. Formic acid, 0.1%

7. 100 mM 2- nitrobenzaldehyde (NBA)

8. Methanol (MeOH) and MilliQ water, 1:1 v/v

Preparation of NP standard solutions

1. **Stock solution of NP Standards (50 ppm) (NP-AOZ, NP-AMOZ, NP-AHD, NP- SEM):** Accurately weigh 1.25 mg ± 0.02mg of each standard and dissolve in 10ml of MeOH in a 25ml volumetric flask and add MeOH to the mark and mix well.

2. **Intermediate Mixture solutions of NP Standards-I (1ppm):** Transfer 460 µl of stock NP AOZ, 332µl of stock NP AMOZ, 432 µl of stock NP-AHD of 554 µl of stock NP – SEM in a 10ml volumetric flask and make upto 10ml with methanol.

3. **Intermediate Mixture solutions of NP Standards-II (100ppb):** Transfer 1000µl of intermediate mixture stock I in a 10ml volumetric flask and add MeOH : MilliQ water (1:1, v/v) to mark. From this, prepare 1, 2, 3, 4 and 5 ppb (S-1 to 5) in Eppendorf tubes.

Preparation of NF metabolites

1. **Stock solution of NF Standards. (50 ppm) AOZ, AMOZ, AHD, SEM:** Accurately weigh 1.25 mg ± 0.02mg of each standard substance and dissolve in 10ml of MeOH in a 25ml volumetric flask and add MeOH to the mark and mix well.

2. **Intermediate Mixture solutions of NF Standards-I (1ppm):** Transfer 460 µl of stock AOZ, 332µl of stock AMOZ, 432 µl of stock AHD of 554 µl of stock SEM in a 10ml volumetric flask and make upto 10ml with methanol.

3. **Intermediate Mixture solutions of NF Standards-II (100ppb):** Transfer 1000µl of intermediate mixture stock I in a 10ml volumetric flask and add MeOH : MilliQ water (1:1, v/v) to mark. From this, prepare 10 (S-10) and 50 ppb (S-50) standards.

Preparation of the sample extract

Take approximately 100g of shrimp sample and homogenize at a low speed for 2 min in a blender. From that, weigh 5± 0.05 gm and add 15ml of HCl (0.2M) and 100mM of 2-nitrobenzaldehyde (0.5ml), vortex for 10 min and place in an orbital shaking incubator at 37°C for 16h. Adjust the pH to 7.0 using 2M NaOH and 1M Na_3PO_4, and add 10ml of ethyl acetate, vortex it again for 10 min. Centrifuge the sample at 5000x g at 4^0C for 10 min. Transfer the upper phase into a clean tube and evaporate the ethyl acetate solution to dryness under the stream of nitrogen at 40°C at 200psi pressure for 20 min. Dissolve the dry residue in 1ml of MeOH: water (5:95). Transfer the supernatant into a syringe and filter it through syringe filter into LC-MS-MS vial.

LC – MS- MS Analysis

Connect the UPLC, LC pump, Auto sampler, and Mass Spectrometer in series. Fix the injection loop on injection port in the auto sampler as 10 μl. Sample transfer tube in AS is 1ml vial. Standard tray contains 48 vials with septa and screw cap in two rows. Fix the UPLC column.

Set the auto sampler method as follows:

Needle height from bottom: 20 Mm

Injection volume and mode: 7.5 μl (Partial loop with needle overfill)

Post-injection valve switch time: 0.0min

Loop loading speed: 5μl/s

Syringe speed: 8μl/s

Flush/Wash source is wash bottle containing 50/50 (v/v) methanol/water.

Flush/Wash volume: 250 μl

Flush speed: 250 μl/s

Tray temperature control: On. Temperature: 15°C

Column Oven Control: On. Temperature: 35°C

Set the method in UPLC as follows:

Mobile phase A: MilliQ Water with 0.1% Formic acid

Mobile phase B: Acetonitrile

Allow mobile phase A/B at 50/50 (v/v) to run through UPLC system for at least 15 min at flow rate 0.3 ml/min to equilibrate the system. Then, set the gradient programme as given below:

Time (min)	Mobile phase, A%	Mobile phase, B%
Initial	95	5
0.50	45	55
3.50	45	55
4.00	95	5
12.00	Controller	Stop

Retention times and molecular weight of the metabolites

Compound	Retention time (min)	MW
NBA-AMOZ	4.37	335
NBA-AH	4.76	249
NBA-AOZ	5.56	236
NBA-SC	5.7	209

Set the TSQ Quantum method as follows:

Acquire time: 12 min

Ion Source: ESI

ESI capillary position: 90 degree

Calibration File Type: High Resolution

Ion Source

Ions spray voltage: 5500V

Ion source Gas 1(GS1) – 50

Ions source Gas 2 (GS 2) – 60

Lurtain Gas (CUR) – 10

Interface heater – ON

Polarity - Positive

Scan type: MRM

Set the MRM Conditions as follows:

Product ID	Parent (m/z) Q1 Mass	Product (m/z) Q3 Mass	Time (Ms)	Collision Energy (v)
SEM -1	209	166	100	16
SEM - 2	209	192	100	19
AOZ-1	236	134	100	30
AOZ -2	236	104	100	18
AHD - 1	249	134	100	30
AHD – 2	249	104	100	17
AMOZ - 1	335	291	100	22
AMOZ -2	335	262	100	27
AMOZ – 3	335	128	100	30

Injection sequence

Set the injection volume as 10 μl

Inject a MeOH blank as the first and second injection

Inject a blank matrix as the third injection

Inject a positive control (Spiked sample with S-10 (10 ppb)) as the fourth injection

Inject S-1 (1 ppb) standard sample both to start and to end the bracket of samples.

Confirmation Criteria

Retention time of the analytes must agree with 5% of that of standard

Two ion ratios of the analytes must agree within 10% of that of standard.

MEDIA COMPOSITION

Alkaline Peptone Salt (APS) Broth

Peptone	-	10.0 g
Sodium chloride	-	30.0 g
Distilled water	-	1000 ml
pH	-	8.5 ±0.2

Sterilize at 121°C for 15 min.

Alkaline Peptone Water (APW)

Peptone	-	10.0 g
Sodium chloride	-	5.0 g
Distilled water	-	1000 ml
pH	-	8.5± 0.2

Sterilize at 121°C for 15 min.

Baird Parker (BP) Medium

Tryptone	-	10.0 g
Beef Extract	-	5.0 g
Yeast Extract	-	1.0 g
Sodium pyruvate	-	10.0 g
Glycine	-	12.0 g
Lithium chloride	-	5.0 g
Agar	-	20.0 g

| Distilled water | - | 1000 ml |
| pH | - | 7.0 ±0.2 |

Sterilize at 121°C for 15 min.

Cool to 50°C and add aseptically add 50 ml concentrated egg yolk emulsion and 3 ml sterile 3.5% potassium tellurite solution. Use the plates within 24 hours of preparation.

Brilliant green lactose bile (BGLB) broth

Peptone	-	10.0 g
Lactose	-	10.0 g
Oxgall or oxbile solution	-	200.0 ml
0.1% brilliant green water solution	-	13.3 ml
Distilled water	-	1000 ml
pH	-	7.2±0.1

Sterilize at 110°C for 10 min.

Butterfield's Buffered Phosphate Diluent

KH_2PO_4	-	26.22 g
Magnesium chloride	-	5 ml
Sodium carbonate	-	7.78 g
Distilled water	-	1000 ml
Final pH	-	7.2 ± 0.1

Sterilize at 121°C for 15 min

Buffered Listeria Enrichment Broth Base (BLEB)

Casein peptone	-	17.0 g
Soy peptone	-	3.0 g
D(+)-dextrose	-	2.5 g
Sodium chloride	-	5.0 g
K_2HPO_4	-	2.5g
Yeast extract	-	6.0 g
Distilled water	-	1000 ml
pH	-	7.3 ± 0.2

Sterilize at 121°C for 15 min

Supplements: Add Cycloheximide - 0.05 g; Nalidixic acid - 0.04 g and Acriflavin HCl - 0.015 g after sterilization

E. coli (EC) Broth

Tryptone	-	20.0 g
Lactose	-	5.0 g
Bile salt Mixture	-	1.5 g
K_2HPO_4	-	4.0 g
KH_2PO_4	-	1.5 g
Sodium chloride	-	5.0 g
Distilled water	-	1000 ml
pH	-	6.9 ± 0.2

Sterilize at 110° C for 10 min.

Eosine Methylene Blue (EMB) Agar

Peptone	-	10.0 g
Lactose	-	5.0 g
Sucrose	-	5.0 g
K_2HPO_4	-	2.0 g
Agar	-	13.5 g
Distilled water	-	1000 ml
Eosine Y	-	0.4 g
Methylene Blue	-	0.065 g
pH	-	7.2± 0.2

Sterilize at 121°C for 15 min..

Fluid Thioglycollate medium

Casein enzymic hydrolysate	-	15.0 g
Yeast extract	-	5.0 g
Dextrose	-	5.5 g
Sodium chloride	-	2.5 g
L-Cystine	-	0.5 g
Sodium thioglycollate	-	0.5 g
Resazurin sodium	-	0.001 g
Water	-	1000 ml
pH	-	7.1± 0.2

Sterilize at 121°C for 15 min.

Fluorescent Pseudomonad Agar

Protease peptone No.3	-	20g
Oxoid Ionagar No. 1	-	12g
Glycerol	-	8 ml
K_2SO_4	-	1.5 g
$MgSO_4$, $7H_2O$	-	1.5 g
Agar	-	15 g
Water	-	950 ml
pH	-	7.2

Sterilize at 121°C for 15 min.

Supplements : Mix penicillin G - 75,000 units, novobiocin – 45 mg and cylcoheximide – 75 mg in 3 ml of 95% ethanol and then dilute to 50 ml with sterile distilled water; and add to 950 ml of melted sterile basal medium at 45°C

Glucose tryptone broth

Tryptone	-	10.0 g
Glucose	-	5.0 g
BCP	-	0.04 g
Water	-	1000 ml
pH	-	6.8-7.0

Sterilize at 121°C for 15 min.

Hektoen enteric (HE) agar

Mixed peptone	-	12.0 g
Yeast extract	-	3.0 g
Sucrose	-	12.0 g
Lactose	-	12.0 g
Salicin	-	2.0 g
Bile salts	-	9.0 g
Sodium chloride	-	5.0 g
Sodium thiosulfate	-	5.0 g
Ammonium ferric citrate	-	1.5 g
Acid fuchsin	-	0.1 g
Bromothymol blue	-	0.065 g

Agar	-	15.0 g
Distilled water	-	1000 ml
pH	-	7.5 ±0.2

Do not autoclave. Sterilize in a boiling water bath for 10-20 min.

Kenner Faecal (KF) Streptococcal agar

Proteose peptone	-	10.0g
Yeast extracts	-	10.0g
Sodium chloride	-	5.0g
Sodium glycerophosphate	-	10.0g
Maltose	-	20.0g
Lactose	-	1.0g
Sodium azide	-	0.4g
Sodium carbonate	-	0.63g
Agar	-	20.0g
Distilled water	-	1000ml
Bromocresol purple	-	0.015g
pH	-	7.2±0.2

Sterilize at 121C for 15 min. Cool to 50C and add 10ml of 1% of 2,3,5,- triphenyl tetrazolium chloride (TTC).

Lactose Broth (LB)

Beef Extract	-	3.0 g
Peptone	-	5.0 g
Lactose	-	5.0 g
Distilled water	-	1000 ml
pH	-	6.9 ±0.1

Sterilize at 121°C for 15 min.

Lauryl Sulphate Tryptose Broth (LSTB)

Tryptone	-	20.0 g
Lactose	-	5.0 g
Sodium chloride	-	5.0 g

K_2HPO_4	-	2.75 g
KH_2PO_4	-	2.75 g
Sodium lauryl sulphate	-	0.1 g
Distilled water	-	1000 ml
pH	-	6.8±0.2

Sterilize at 110°C for 10 min.

PALCAM Agar

Peptone	-	23.0 g
Starch	-	1.0 g
Lithium chloride	-	15.0 g
Aesculin	-	0.8 g
Ammonium Ferric citrate	-	0.5 g
Sodium chloride	-	5.0 g
Mannitol	-	10.0 g
Dextrose	-	0.5 g
Phenol red	-	0.08 g
Agar	-	13.0 g
Distilled water	-	1000.0 ml
pH	-	7.0± 0.2

Sterilize at 121°C for 15 min.

Supplements: Add Polymyxin B sulfate- 0.01 g; Ceftazidime-0.02 g ; and Acriflavin HCl -0.005 g after sterilization.

PE-2 Medium

Yeast extract	-	3.0 g
Peptone	-	20 g
BCP (2%)	-	2.0 ml
Water	-	1000 ml
Green peas	- 4-5/ tube	

Sterilize at 121°C for 30 min.

Peptone Iron Agar (PIA)

Peptone	-	15.0 g
Proteose peptone	-	5.0 g
Ferric ammonium citrate	-	0.5 g
Sodium glycerophosphate	-	1.0 g
Sodium thiosulphate	-	0.08 g
Agar	-	15.0 g
Distilled water	-	1000 ml
pH	-	6.7±0.2

Sterilize at 121°C for 15 min

Plate Count Agar (PCA)

Tryptone	-	5.0 g
Yeast Extract	-	2.5 g
Dextrose	-	1.0 g
Agar	-	15.0 g
Distilled water	-	1000 ml
Final pH	-	7.0 ±0.2

Sterilize at 121°C for 15 min.

Plate count agar with antibiotics or tartaric acid

Tryptone	-	5.0 g
Yeast extract	-	2.5 g
Dextrose	-	1.0 g
Agar	-	15.0 g
Distilled water	-	1000 ml
pH	-	7.0±0.2

Sterilize at 121°C for 15 min and cool to 50°C.

Add antibiotics (chlorotetracycline-HCl) at 40 ppm or tartaric acid, 10% and mix well.

Potato dextrose agar (PDA)

Potato infusion	-	200 ml
Dextrose	-	20.0 g

Agar	-	20.0 g
Distilled water	-	1000 ml
pH	-	5.4±0.2

Sterile at 121°C for 15 min

Pseudomonas Agar F

Tryptone	-	10 g
Protease peptone No. 3	-	10 g
K_2HPO_4	-	1.5 g
$MgSO_4$, $7H_2O$	-	1.5 g
Agar	-	15 g
Water	-	1000 ml
pH	-	7.0

Sterilize at 121°C for 15 min.

Pseudomonas Isolation Agar

Peptone	-	20 g
$MgCl_2$	-	1.4g
K_2SO_4	-	1.5g
Irgasan	-	0.025g
Agar	-	15g
Water	-	1000 ml
pH	-	7.0

Sterilize at 121°C for 15 min.

Rappaport-Vassiliadis (RV) medium

Papaic digest of soyabean meal	–	4.50 g
Sodium chloride	-	7.20 g
Monopotassium phosphate	-	1.44 g
Magnesium chloride	-	36.00 g
Malachite green	-	0.036 g
Distilled water	-	1000 ml
pH	-	5.2 ± 0.2

Sterilize at 115°C for 15 min.

Selenite Cystine Broth (SCB)

Peptone	-	5.0 g
Lactose	-	4.0 g
Sodium biselenite	-	4.0 g
Disodium hydrogen phosphate	-	10.0 g
L-Cystine	-	0.01 g
Distilled water	-	1000 ml
pH	-	7.0 ± 0.2

Do not autoclave. Sterilize in a boiling water bath for 10-20 min.

Tetrathionate Broth (TTB)

Peptone	-	18.0 g
Calcium carbonate	-	25.0 g
Sodium thiosulphate	-	38.0 g
Yeast extract	-	2.0 g
Sodium chloride	-	5.0 g
D-Mannitol	-	2.5 g
Dextrose	-	0.5 g
Sodium deoxycholate	-	0.5 g
Brilliant green	-	0.01 g
Distilled water	-	1000.0 ml
pH	-	7 ± 0.2

Do not autoclave. Sterilize in a boiling water bath for 10-20 min. Cool to 45°C. Add 40 ml of iodine solution (8 g potassium iodide and 5 g iodine dissolved per 40 ml distilled water) mix on the day of use.

Thiosulphate Citrate Bile salt Sucrose (TCBS) Agar

Yeast extract	-	5.0 g
Peptone	-	10.0 g
Sucrose	-	20.0 g
Sodium thiosulphate	-	10.0 g
Sodium citrate dihydrate	-	10.0 g
Sodium cholate	-	3.0 g

Oxgall	-	5.0 g
Sodium chloride	-	10.0 g
Ferric citrate	-	1.0 g
Bromothymol blue (BTB)	-	0.04 g
Thymol blue (TB)	-	0.04 g
Agar	-	15.0 g
Distilled water	-	976.0 ml
pH	-	8.6

Dissolve and boil the ingredients except BTB and TB. Add 20 ml of 0.2% BTB and 4 ml of 1% TB. Heat again to boil and cool to 45°C. Do not autoclave.

Tryptic Soy Agar (TSA)

Tryptone	-	17 g
Soytone	-	3 g
NaCl	-	5 g
K_2HPO_4	-	1.5 g
Agar	-	15 g
pH	-	7.3

Sterilize at 121°C for 15 min.

Violet red bile agar (VRBA)

Peptic digest of meat	-	7.0 g
Yeast extract	-	3.0 g
Lactose	-	10.0 g
Bile salts	-	1.5 g
Sodium chloride	-	5.0 g
Neutral red	-	30.0 mg
Crystal violet	-	2.0 mg
Agar	--	12.0 g
Distilled water	-	1000 ml
pH	-	7.4 ± 0.2

Sterilize at 121°C for 15 min.

Wagatsuma Agar

Yeast extract	-	3.0 g
Peptone	-	10.0 g
Sodium chloride	-	70.0 g
K_2HPO_4	-	5.0 g
Mannitol	-	10.0 g
Crystal violet	-	0.0001 g
Agar	-	15.0 g
Distilled water	-	1000.0 ml
pH	-	8.0

Do not autoclave. Steam at 100°C for 30 min. Add 100 ml of 20 % suspension of freshly drawn citrated human red blood cells or rabbit RBCs washed three times with sterile physiological saline. Mix well and pour into sterile dishes and cool.

Xylose-Lysine Deoxycholate (XLD) Agar

Xylose	-	3.5 g
L-Lysine	-	5.0 g
Lactose	-	7.5 g
Sucrose	-	7.5 g
Yeast extract	-	3.0 g
Sodium chloride	-	5.0 g
Sodium deoxycholate	-	2.5 g
Sodium thiosulphate	-	6.8 g
Ferric Ammonium citrate	-	0.8 g
Phenol red	-	0.08 g
Agar	-	15.0 g
Distilled water	-	1000 ml
pH	-	7.4 ±0.2

Do not autoclave. Sterilize in a boiling water bath for 10-20 min.

Bismuth Sulphite Agar

Peptone	-	10.0 g
HM Peptone B	-	5.0 g
Dextrose (Glucose)	-	5.0 g

Disodium phosphate	-	4.0 g
Ferrous sulphate	-	0.30 g
Bismuth sulphite indicator	-	8.0 g
Brilliant green	-	0.025 g
Agar	-	20.0 g
Water	-	1000 ml
Final pH (at 25°C)	-	7.7±0.2

Do not sterilize in autoclave. Heat to boiling to dissolve the medium completely.

Tergitol-7 Agar

Peptic digest of animal tissue	-	10.0 g
Yeast extract	-	6.0 g
Meat extract	-	5.0 g
Lactose	-	20.0 g
Sodium heptadecyl sulphate (Tergitol 7)-		0.10 g
Bromo thymol blue	-	0.050 g
Agar	-	16.0 g
Water	-	1000 ml
Final pH (at 25°C)	-	7.2±0.2

Sterilize by autoclaving at 121°C for 15 minutes. Cool to 45-50°C. Add 2.5 ml of 1% Triphenyl Tetrazolium Chloride (TTC).

Oxford Agar

Peptone, special	-	23.0 g
Lithium Chloride	-	15.0 g
Sodium chloride	-	5.0 g
Corn Starch	-	1.0 g
Esculin	-	1.0 g
Ammonium Ferric Citrate	-	0.50 g
Agar	-	12.0 g
Water	-	1000 ml
Final pH (at 25°C)	-	77.2±0.1

Sterilize by autoclaving at 121°C for 15 minutes. Cool to 45-50°C and aseptically add the rehydrated contents of 1 vial of Listeria Moxalactam Supplement, Modified (FD126F).

Listeria Moxalactam Supplement

Colistin	-	10.0 mg
Moxalactam	-	15.0 mg

Agar Listeria according to Ottaviani and Agosti (ALOA)

HM Peptone	-	18.0 g
Tryptone	-	6.0 g
Yeast extract	-	10.0 g
Sodium pyruvate	-	2.0 g
Dextrose (Glucose)	-	2.0 g
Magnesium glycerophosphate	-	1.0 g
Magnesium sulphate	-	0.50 g
Sodium chloride	-	5.0 g
Lithium chloride	-	10.0 g
Disodium hydrogen phosphate anhydrous	-	2.50 g
5-Bromo-4 chloro-3-indolyl- β D-glucopyranoside	-	0.050 g
Agar	-	15.0 g
Water	-	1000 ml
Final pH (at 25°C)	-	7.2±0.2

Sterilize by autoclaving at 121°C for 15 minutes. Cool to 45-50°C. Aseptically add sterile contents of 1 vial of L. mono Enrichment Supplement I (FD214) and sterile rehydrated contents of L. mono Selective Supplement I (FD212), L .mono Selective Supplement II (FD213).

L. mono Enrichment Supplement I FD214 (Per vial sufficient for 500 ml medium)

L – phosphatidylinositol	-	1.0 g
Distilled water	-	25.0 ml

L. mono Selective Supplement I (FD212) (Per vial sufficient for 500 ml medium)

Polymyxin B sulphate	-	38350IU

Rehydrate the contents of 1 vial aseptically with 10 ml sterile distilled water.

L .mono Selective Supplement II (FD213) (Per vial sufficient for 500 ml medium)

Ceftazidime	-	10.0 mg
Amphotericin B	-	5.0 mg
Nalidixic acid, sodium salt	-	10.0 mg

Rehydrate the contents of 1 vial aseptically with 2 ml of 0.2 N Sodium hydroxide, further add 3 ml of sterile distilled water.

Lithium chloride phenylethanol moxalactam (LPM) Agar

Casein enzymic hydrolysate	-	5.0 g
Peptic digest of animal tissue	-	5.0 g
Beef extract	-	3.0 g
Glycine anhydride	-	10.0 g
Lithium chloride	-	5.0 g
Sodium chloride	-	5.0 g
Phenylethyl alcohol	-	2.50 g
Agar	-	15.0 g
Water	-	1000 ml
Final pH (at 25°C)	-	7.3±0.2

Sterilize by autoclaving at 121°C for 12 minutes. Cool to 50°C and aseptically add rehydrated contents of 1 vial of Moxalactam Supplement (FD151).

Moxalactam Supplement (FD151) (sufficient for 1000 ml medium)

Moxalactam	-	20.0 mg

Rehydrate the contents of one vial aseptically with 10 ml of sterile distilled water.

Differential Reinforced Clostridial Medium (DRCM)

Casein enzymic hydrolysate	-	5.0 g
Peptic digest of animal tissue	-	5.0 g
Beef extract	-	8.0 g
Yeast extract	-	1.0 g
Starch	-	1.0 g
Sodium acetate	-	5.0 g
Glucose	-	1.0 g
L-Cysteine hydrochloride	-	0.50 g
Sodium bisulphite	-	0.50 g
Ferric ammonium citrate	-	0.50 g
Resazurin	-	0.0020 g
Agar	-	15.0 g
Water	-	1000 ml
Final pH (at 25°C)	-	7.1±0.2

Sterilize by autoclaving at 121°C for 15 minutes.

Buffered Peptone Water with NaCl

Peptone	-	1.0 g
Potassium dihydrogen phosphate	-	3.56 g
Disodium hydrogen phosphate	-	7.23 g
Sodium chloride	-	4.30 g
Water	-	1000 ml
Final pH (at 25°C)	-	7.0±0.2

Heat, if necessary, to dissolve the medium completely. Add 0.1 to 1% w/v polysorbate 20 or 80, if desired. Sterilize by autoclaving at 121°C for 15 minutes.

Fraser Broth

Enzymatic digest of animal tissues	-	5.0 g
Enzymatic digest of casein	-	5.0 g
Yeast extract	-	5.0 g
Meat extract	-	5.0 g
Sodium chloride	-	20.0 g
Disodium hydrogen phosphate dihydrate	-	12.0 g
Potassium dihydrogen phosphate	-	1.35 g
Esculin	-	1.0 g
Lithium chloride	-	3.0 g
Water	-	1000 ml
Final pH (at 25°C)	-	7.2±0.2

Sterilize by autoclaving at 121°C for 15 minutes. Add the following supplements after sterilization.

Supplements

	Half Fraser	Fraser
Acriflavin hydrochloride	0.0125 g	0.025 g
Nalidixic acid, sodium salt	0.01 g	0.02 g
Ammonium Iron citrate	0.50 g	0.50 g

Phosphate Buffered Saline

Disodium hydrogen phosphate, anhydrous	-	0.795 g
Potassium dihydrogen phosphate	-	0.144 g
Sodium chloride	-	9. 0 g
Water	-	1000 ml

Rappaport-Vassiliadis Soya (RVS) Broth

Papaic digest of soyabean meal	-	4.50 g
Sodium chloride	-	8.0 g
Potassium dihydrogen phosphate	-	0.60 g
Dipotassium phosphate	-	0.40 g
Magnesium chloride, hexahydrate	-	29.0 g
Malachite green	-	0.036 g
Water	-	1000 ml
Final pH (at 25°C)	-	5.2±0.2

Sterilize by autoclaving at 121°C for 15 minutes.

Lactobacillus MRS Agar

HM extract B	-	8.0 g
Peptone	-	10.0 g
Yeast extract	-	5.0 g
Ammonium citrate	-	2.0 g
Sodium acetate	-	5.0 g
Magnesium sulphate, heptahydrate	-	0.20 g
Manganese sulphate, tetrahydrate	-	0.050 g
Dipotassium phosphate	-	2.0 g
Glucose, anhydrous	-	20.0 g
Polysorbate 80 (Tween 80)	-	1.0 ml
Agar	-	12.0 g
Water	-	1000 ml
Final pH (at 25°C)	-	5.7±0.2

Sterilize by autoclaving at 121°C for 15 minutes.

Mueller-Kauffman Tetrathionate Novobiocin (MKTTn) Broth Base

Meat extract	-	4.30 g
Enzymatic digest of casein	-	8.60 g
Ox bile for bacteriological use	-	4.78 g
Sodium chloride	-	2.60 g
Calcium carbonate	-	38.70 g

Sodium thiosulphate, pentahydrate	-	47.80 g
Brilliant green	-	0.0096 g
Water	-	1000 ml
Final pH (at 25°C)	-	8.0±0.2

Do not autoclave. Heat the medium just to boiling. Cool to 45-50°C and just before use aseptically add rehydrated contents of 1 vial of MKTT Novobiocin Supplement (FD203) and 20 ml of iodine-iodide solution (20 g iodine and 25 g potassium iodide in 100 ml sterile distilled water). Mix well to disperse calcium carbonate uniformly before dispensing in sterile tubes.

Supplements

Novobiocin sodium salt	-	0.04 g
Iodine-iodide solution	-	20.0 ml (20 g iodine and 25 g potassium iodide in 100 ml sterile distilled water)

Salt Polymyxin Broth Base

Tryptone	-	10.0 g
Yeast extract	-	3.0 g
Sodium chloride	-	20.0 g
Water	-	1000 ml
Final pH (at 25°C)	-	8.8±0.2

Sterilize by autoclaving at 121°C for 15 minutes. Cool to 45 - 50°C and aseptically add rehydrated contents of 1 vial of Polymyxin B Selective Supplement (FD003). Mix well and dispense into sterile tubes or flasks as desired.

Selective Supplement

Polymyxin B sulphate	-	50000Unit

Rehydrate the contents of one vial aseptically with 2 ml sterile distilled water. Mix well and aseptically add it to 500 ml of agar

Tryptone Sucrose Tetrazolium Agar Base (TSTA)

Casein enzymic hydrolysate	-	15.0 g
Papaic digest of soyabean meal	-	5.0 g
Sodium chloride	-	30.0 g
Saccharose	-	20.0 g
Bile salts	-	0.50 g
Agar	-	15.0 g

| Water | - | 1000 ml |
| Final pH (at 25°C) | - | 7.1±0.2 |

Sterilize by autoclaving at 121°C for 15 minutes. Cool to 45-50°C and aseptically add 3 ml of 1% 2, 3, 5-Triphenyl Tetrazolium Chloride (TTC).

Yeast extract dextrose chloramphenicol agar

Yeast extract	-	5.0 g
Dextrose (Glucose)	-	20.0 g
Chloramphenicol	-	0.10 g
Agar	-	14.90 g
Water	-	1000 ml
Final pH (at 25°C)	-	6.6±0.2

Sterilize by autoclaving at 121°C for 15 minutes.

Modified Listeria Oxford Agar Base

Peptone special	-	23.0 g
Corn starch	-	1.0 g
Sodium chloride	-	5.0 g
Aesculin	-	1.0 g
Iron (III) Ammonium citrate	-	0.50 g
Lithium chloride	-	12.0 g
Agar	-	10.0 g
Water	-	1000 ml
Final pH (at 25°C)	-	7.2±0.2

Sterilize by autoclaving at 121°C for 15 minutes. Cool to 45-50°C and aseptically add the rehydrated contents of 1 vial of Modified Listeria Oxford Selective Supplement.

Modified Listeria Oxford Selective Supplement (for 1000 ml)

| Colistin sulfate | - | 10 mg |
| Ceftazidime pentahydrate | - | 20 mg |

Rehydrate the contents of 1 vial aseptically with 10 ml sterile distilled water. Mix well and aseptically add to 1000 ml of sterile, molten, cooled (45-50°C) Modified Listeria Oxford Agar Base

Vibrio vulnificus agar

Solution 1

Peptone	-	20.0 g
NaCl	-	3.0 g
Bromothymol blue*	-	10.0 ml
Agar	-	25.0 g
Water	-	900 ml
pH	-	8.2±0.2

Sterilize by autoclaving at 121°C for 15 minutes.

*Dissolve 0.6 g of Bromothymol blue in 100 ml of 70% ethanol.

Supplement (Solution 2)

Cellobiose	-	10.0 g
Distilled water	-	100 ml

Dissolve cellobiose in distilled water by heating gently. Cool and filter sterilize.

Add Solution 2 to cooled Solution 1, mix, and dispense into petri dishes. Final colour is light blue.

REFERENCES

Agersborg, A. Dahl, R. and Martinez, I. (1997). Sample preparation and DNA extraction procedures for polymerase chain reaction identification of Listeria monocytogenes in seafoods. *Int. J. Food Microbiol.* 35: 275-280.

Aitken, A. and Learmonth, M.P. (2002). Protein determination by UV absorption. In: *The Protein Protocols Handbook.* 2nd edn. Ed. Walker JM, Human Press Inc., Totowa, NJ.

Andrews, W.H., Wang, H., Jacobson and Hammack, T. (2019). Chapter 5: Salmonella, In: Bacteriological Analytical Manual. 8th edn. Gaithersburg, MD, US Food and Drug Administration, December.

AOAC. (1995). Method Cd. 18-90. In: *Official Methods and Recommended Practices of the Americal Oil Chemist' s Society.* Firestone D Ed. AOCS, Champaign, IL.

AOAC. (1990). Official methods. 960.39. In: *Official Methods of Analysis.* 15th edn. Association of Official Analytical Chemists, Arlington, VA.

AOAC. (1990). Official Methods of Analysis (AOAC, 948.22.). Lipid extraction by Soxhlet method. Association of Official Analytical Chemists, 15th Ed. Arlington, VA.

AOAC. (1995). Estimation of Phosphorous. AOAC, 986.24. Official Methods of Analysis of AOAC International, 16th edn, AOAC International, Arlington, Virginia, USA.

AOAC. (1995). Minerals in Seafood. AOAC, 938.28. Official Methods of Analysis of AOAC International, 16th edn, AOAC International, Arlington, Virginia, USA.

AOCS. (1995). Method Cd 8-53. In: *Official Methods and Recommended Practices of the American Oil Chemist' s Society*. Firestone D Ed. AOCS, Champaign, IL.

Aquaculture Products: Detection, identification and confirmation of Residues of Nitrofuran Metabolites by LC MS MS EU Reference method from State Institute for Quality Control of Agriculture Products – (RIKLIT-2002), Bornsesteeg 45, P O Box 230, Netherlands 6700 AE. National Referral Laboratory. LC MSMS DOC:SOP/NF/00, Issue date: 17-02-2014.

Atwater, W.O. and Woods, C.D. (1896). The chemical composition of American food materials. United States Dept. of Agriculture Office of Experiment Stations, Bulletin 28, Washington DC, Government Printing Office.

Aubourg, S.P. and Medina, I. (1999). Influence of storage time and temperature on lipid deterioration during cod (*Gadus morhua*) and haddock (*Melanogrammus aeglefinus*) frozen storage. *J. Sci. Food Agri*, 79, 1943-1948.

Blackstone, G.M., Nordstrom, J.L., Bowen, M.D., Meyer, R.F., Imbro, P. and DePaola, A. (2007). Use of a real time PCR assay for detection of the ctxA gene of Vibrio cholerae in an environment survey of Mobile Bay. *J. Microbiol. Methods* 68: 254-259.

Bligh, E.G. and Dyer, W.J. (1959) A rapid method of total lipid extraction and purification. *Can. J. Biochem. Physiol.* 37, 911-917.

Bradford, M.A. (1976). Rapid and sensitive method for the quan titation of microgram quantities of protein utilizing the principle of protein-dye binding, *Anal. Biochem.* 72: 248–254.

Branch, A.C. and Vail, A.MA. (1985). Bringing fish inspection into the computer age. *Food Technol. Aust.* 37(8): 352-355.

Brasher, C.W., DePaola, A., Jones, D.D. and Bej, A.K. (1998). Detection of microbial pathogens in shellfish with multiplex PCR. *Curr. Microbiol.* 37: 101-107.

Bremner, H.A. (1985). A convenient, easy-to-use system for estimating the quality of chilled seafood. *Fish Processing Bulletin.* 7: 59-70.

Carpenter, K.J. (1960). The estimation of the available lysine in animal-protein foods. *Biochem. J.* 77: 604.

Carroll, N.V., Longley, R.W. and Roe, J.H. (1955). The determination of glycogen in liver and muscle by use of anthrone reagent. Dept. of Biochemistry, School of Medicine, George Washington University, Washington, D.C., *J. Biol. Chem.*, 583-593.

Chawla, S.P., Venugopol, V., and Nair, P.M. (1996). Gelation of proteins from washed muscle of threadfin bream (*Nemipterus japonicus*) under mild acidic conditions. *J. Food Science,* 61, 362–371.

Cheow, C.S., Norizah, M. S., Kyaw, Z. Y., and Howell, N.K. (2007). Preparation and characterization of gelatin from skin of sin croaker (*Johnius dussumieri*) and short scad (*Decapterus macrosoma*). *Food Chemistry,* 101, 386-391.

Cheow, C.S., Kyaw, Z.Y., Howell, N.K. and Dzulkifly, M.H. (2004). Relationship between physicochemical properties of starches and expansion of fish cracker 'keropok'. *Journal of Food Quality,* 27(1):1-12.

Cho, S.H., Jahncke, M.L., Chin, K.B., and Eun, J.B. (2006). The effect of processing conditions on the properties of gelatin from skate (*Raja kenojei*) skins. *Food Hydrocolloids,* 20, 810-816.

Cho, S.M., Kwak, K.S., Park, D.C., Gu, Y.S., Ji, C.I., Jang, D.H., *et al.* (2004). Processing optimization and functional properties of gelatin from shark (*Isurus oxyrinchus*) cartilage. *Food Hydrocolloids,* 18, 573-579.

Christie, R., Kent, M. and Lees, A. (1985) Microwave and infra-red drying versus conventional oven drying methods for moisture determination in fish flesh. *Int. J. Food Sci. Technol.* 20(2), 117-127.

Coastal Aquaculture Authority (CAA) of India Notification (2005). List of Antibiotics and other pharmacologically active substances banned for using in shrimp aquaculture.

Commission Recommendation 2013/711/EU of 3 December 2013 on the reduction of the presence of dioxins, furans and PCBs in feed and food as amended by Commission Recommendation 2014/663/EU of 11 September 2014.

Commission Regulation (EC) No. 1881/2006 of 19 December 2006 setting maximum levels for certain contaminants in foodstuffs.

Commission Regulation (EU) No. 37/2010 of 22 December 2009 on pharmacologically active substances and their classification regarding maximum residue limits in foodstuffs of animal origin.

Detection, identification and confirmation of Chloramphenicol by LC MS MS. EU Reference method from Agence Francaise de Securite Sanitaire De Ailments (AFSSA). Laboratory for Veterinary medicine, FOUGERES, Cedex, Javene – 351133 FRANCE, European Union Referral. LC MSMS, DOC: SOP/CAP/00, Issue date: 09-06-2014

Dubois, M., Gilles, K.A., Hamilton, J.K., Rebers, P.A. and Smith, F. (1956) Colorimetric method for determination of sugars and related substances. *Anal. Chem.* 18, 350-356.

Dumas, J.B.A. (1831). Procedes de l'analyseorganique. *Ann. Chem. Phys.*, 247, 198-213.

Dunn, M.J. (1992). Protein determination of total protein concentration, In Harris, E.L.V., Angal S. Eds, Protein Purification Methods, Oxford: IRL Press.

EPS. (1994). Determination of Mercury in Water by Cold Vapour Atomic Absorption Spectrometry, EPA Method 245. 1, Environmental Monitoring Systems Laboratory, Revision 3.0, May 1994.

Feng, P., Weagent, S.D., Grant, M.A. and Burkhardt, W. (2017). Chapter 4. Enumeration of *Escherichia coli* and the Coliform Bacteria. Bacteriological Analytical Manual, Gaithersgurg, MD, US Food and Drug Administration.

Folch, J., Lees, M. and Stanley, G.H.S. (1957). Preparation of lipids extracts from brain tissue. *J. Biol. Chem.* 226,497-509.

Folch, J., Lees, M. and Stanley, G.S. (1957). A simple method for the isolation and purification of total lipids from animal tissues. *Journal of biological chemistry*, 226(1):497-509.

FSSAI (2019). Direction under section 16(5) of Food Safety and Standards Act, 2006 on adhoc-limit of fermaldehyde in fish and fisheries products dated 10.06.2019.

Greenfield, H. and Southgate, D.A.T. (2003). Food composition data, production, management and use. Food and Agricultural Organization of the United Nations, Rome.

Greisen, K., Loeffelholz, M., Purohit, A. and Leong, D. (1994) PCR primers and probes for the 16S rRNA gene of most species of pathogenic bacteria, including bacteria found in cerebrospinal fluids. *J. Clin. Microbiol.* 32: 335-351.

Harlin, K., Surratt, K., and Peters, C. (1995). Standard Operating Procedure for the Analysis of PCBs and Organochlorine Pesticides by GC-ECD. Volume 2, WB. Chapter 1. Standard Operating Procedure CH-IN-002.3.

Heems, D., Luck, G., Fraudeau, C. and Verette, E. (1998). Fully automated precolumn derivatization, on line dialysis and high-performance liquid chromatographic analysis of amino acids in food, beverages and feedstuff. *J. Chrom.* Part A., 798, 9-17.

Hill, W.E., Keasler, S.P., Trucksess, M.W., Feng, P., Kaysner, C.A. and Lampel, K.A. (1991). Polymerase chain reaction identification of Vibrio vulnificus in artificially contaminated oysters. *Appl. Environ. Microbiol.* 57: 707-711.

Hitchins, A.D., Jinneman, K. and Chen. Y. (2017) Chapter 10. Detection of *Listeria monocytogenes* in Foods and Environmental Samples and Enumeration of *Listeria monocytogenes* in foods. In: Bacteriological Analytical Manual, Gaithersgurg, MD, US Food and Drug Administration.

Huang, WB. (2003). Heavy metal concentrations in the common benthic fishes caught from the coastal waters of eastern Taiwan. Fenxi, 11 (4): 324-330.

International Union of Pure and Applied Chemistry (IUPAC) (1987). *Standard Methods for the Analysis of Oils, Fats and Derivatives*, Paquot C and Hautfenne A (Eds.), 7th Revised Version, Blackwell Scientific, London.

IS 14988 (Part 1): 2001 / ISO 11290- 1: 1996. Microbiology of food and animal feeding stuffs - Horizontal method for detection and enumeration of Listeria monocytogenes : Part 1 Detection method.

IS 14988 (Part 2): 2002 / ISO 11290- 2: 1998. Microbiology of food and animal feeding stuffs - Horizontal method for detection and enumeration of Listeria monocytogenes : Part 2 Enumeration.

IS 5402:2002 / ISO 4833:1991. General guidance for the enumeration of microorganisms - Colony count technique at 30 degree C (first revision).

IS 5403: 1999. Method for Yeast and Mould Count of Foodstuffs and animal feeds.

IS 5887 (Part 8/Sec 1): 2002 / ISO 6888- 1: 1999. Methods for detection of bacteria responsible for food poisoning: Part 8 Horizontal method for enumeration of coagulase-positive staphylococci (Staphylococcus Aureus and other species) Section 1 Technique using Baird-Parker Agar Medium.

IS 5887 (Part 8/Sec 2): 2002 / ISO 6888- 2: 1999. Methods for detection of bacteria responsible for food poisoning: Part 8 Horizontal method for enumeration of coagulase-positive staphylococci (Staphylococcus Aureus and other species) Section 2 Technique using rabbit plasma fibrinogen.

ISO 11290-1: 2017 - Microbiology of the food chain - Horizontal method for the detection and enumeration of Listeria monocytogenes and of Listeria spp. - Part 1: Detection method, International Organization for Standardization , Geneva, Switzerland.

ISO 11290-2: 2017 - Microbiology of the food chain — Horizontal method for the detection and enumeration of Listeria monocytogenes and of Listeria spp. — Part 2: Enumeration method, International Organization for Standardization, Geneva, Switzerland.

ISO 16649-2: 2001. Microbiology of food and animal feeding stuffs — Horizontal method for the enumeration of beta-glucuronidase-positive Escherichia coli — Part 2: Colony-count technique at 44 degrees C using 5-bromo-4-chloro-3-indolyl beta-D-glucuronide, International Organization for Standardization, Geneva, Switzerland.

ISO 21871: 2006. Microbiology of Food and Animal Feeding Stuffs: Horizontal Method for the Detection of Potentially Enteropathogenic *Vibrio* spp. Part 2: Detection of Species other than Vibrio parahaemolyticus and Vibrio cholerae, Geneva, Switzerland.

ISO 21872-1: 2017. (2017). Microbiology of the food chain — Horizontal method for the determination of *Vibrio* spp. — Part 1: Detection of potentially enteropathogenic *Vibrio parahaemolyticus, Vibrio cholerae* and *Vibrio vulnificus*. The International Organization for Standardization, Geneva, Switzerland.

ISO 4833-1: 2013. Microbiology of the food chain — Horizontal method for the enumeration of microorganisms — Part 1: Colony count at 30°C by the pour plate technique, International Organization for Standardization , Geneva, Switzerland.

ISO 6579-1: 2017. Microbiology of the food chain — Horizontal method for the detection, enumeration and serotyping of Salmonella — Part 1: Detection of Salmonella spp. International Organization for Standardization , Geneva, Switzerland.

ISO 6888-1:1999/AMD 1: 2003. Microbiology of food and animal feeding stuffs — Horizontal method for the enumeration of coagulase-positive staphylococci (Staphylococcus aureus and other species) — Part 1: Technique using Baird-Parker agar medium — Amendment 1: Inclusion of precision data, International Organization for Standardization , Geneva, Switzerland.

ISO 6888-1:1999/AMD 2: 2018. Microbiology of food and animal feeding stuffs — Horizontal method for the enumeration of coagulase-positive staphylococci (Staphylococcus aureus and other species) — Part 1: Technique using Baird-Parker agar medium — Amendment 2: Inclusion of an alternative confirmation test using RPFA stab method. International Organization for Standardization, Geneva, Switzerland.

ISO 6888-2: 1999. Microbiology of food and animal feeding stuffs — Horizontal method for the enumeration of coagulase-positive staphylococci (Staphylococcus aureus and other species) — Part 2: Technique using rabbit plasma fibrinogen agar medium. International Organization for Standardization, Geneva, Switzerland.

ISO/TS 17919: 2013. Microbiology of the food chain — Polymerase chain reaction (PCR) for the detection of food-borne pathogens — Detection of botulinum type A, B, E and F neurotoxin-producing clostridia.

Jeyasekaran, G., Jeevithan, E., Jeyashakila, R., Shalini, R., Thirumalairaj, K. and Thangarani, A.J. (2014). Multiplex PCR assay for rapid detection of different strains of *Escherichia coli* associated with fish and fishery products. *RRJoLS*. 4:1-9.

Jeyasekaran, G., Thirumalairaj, K., Jeyashakila, R., Thangarani, J.A., Karthika, S. and Luci, M. (2011). Simultaneous detection of *Staphylococcus aureus* enterotoxin C-producing strains from clinical and environmental samples by multiplex PCR assay. *Ann. Microbiol.* 61: 585-590.

Jeyasekaran, G., Thirumalairaj, K., Jeyashakila, R., Thangarani, A.J. and Sukumar, D. (2012). Detection of *Salmonella enterica* serovars in shrimps in eight hours by multiplex PCR assay. *Ann. Microbiol.* 62: 225-231.

Jeyasekaran, G., Thirumalairaj, K., Jeyashakila, R., Thangarani, A.J. and Sukumar, D. (2011). Multiplex polymerase chain reaction-based assay for the specific detection of toxin-producing *Vibrio choleare* in fish and fishery products. *Appl. Microbiol. Biotechnol.* 90: 1111-1118.

Jiang, S.T., Hwang, B.S. and Tsao, C.Y. (1987). Protein denaturation and changes in nucleotides of fish muscle during frozen storage, *J. Agri. Food Chem.,* 35: 22-27.

Jonsdottir, S. (1992). Quality index method and TQM system. In: Quality issues in the fish industry (Eds. R. Olafsson and A.H. Ingthorsson). Research Liaison Office, University of Iceland, Reykjavik, Iceland. pp. 81-94.

Kaysner, C.A. and DePaola, A. (2004). *Vibrio cholerae, V. parahaemolyticus, V. vulnificus* and other *Vibrio* spp. In: Bacteriological Analytical Manual. 8th edn. Chapter 9: Revision A, Gaithersburg MD., US Food and Drug Administration.

Ke, P.J. and Woyewoda, A.D. (1979) Microdetermination of thiobarbituric acid values in marine lipids by a direct spectrophotometric method with a monophasic reaction system. *Ana. Chim. Acta.,* 106, 279-284.

Kent, M. (1990) Measurement of dielectric properties of herring flesh using transmission time domain spectroscopy. *Int. J. Food Sci. Technol.*, 25, 26-38.

Kim, Y.B., Okuda, J., Matsumoto, C., Takahashi, N., Hashimoto, S. and Nishibuchi, M. (1999) Identification of Vibrio parahaemolyticus strains at the species level by PCR targeted to the toxR gene. *J. Clin. Microbiol.* 37: 1173-1177.

Kjeldahl, J. (1883) A new method for the determination of nitrogen in organic matter. *Z. Anal. Chem.*, 22, 366-382.

Koch, W.H., Payne, W.L., Wentz, B.A. and Cebula, T.A. (1993) Rapid polymerase chain reaction method for detection of Vibrio cholerae in foods, *Appl. Environ. Microbiol.* February, 556-560.

Kumar, H.S., Parvathi, A., Karunasagar, I. and Karunasagar, I. (2006) A gyr B based PCR for the detection of *Vibrio vulnificus* and its application for direct detection of this pathogen in oyster enrichment broths. *Int. J. Food Microbiol.* 111: 216-220.

Laemmli, U.K. (1970). Cleavage of structural proteins during the assembly of the head of bacteriophage T4. *Nature*, 227(5259): 680-685.

Lee, J. and Levin, R. (2006a) Selection of universal primers for PCR quantification of total bacteria associated with fish fillets. *Food Biotechnol.* 20: 275-286.

Lee, J. and Levin, R. (2006b). Direct application of the polymerase chain reaction for quantification of total bacteria on fish fillets. *Food Biotechnol.* 20: 287-298.

Lee, J. and Levin, R. (2007). Rapid quantification of total bacteria on cod fillets by using real time PCR. J. Fisheries Science, 1: 58-67.

Linda, M.N. Palm, Derick Carboo, Philip O. Yeboah, Winston J. Quasie, Mordecai A. Gorleku and Albert Darko. (2011) Characterization of Polycyclic Aromatic Hydrocarbons (PAHs) Present inSmoked Fish from Ghana Advance Journal of Food Science and Technology 3(5): 332-338.

Lowry, O.H., Rosebrough, N.J., Farr, A.L. and Randall, R.J. (1951) Protein measurements with the Folin phenol reagent. *J. Biol. Chem.*, 193, 263-275.

Martinsdóttir, E., Sveinsdottir, K., Luten, J., Schelvis-Smit, R., and Hyldig, G. (2001). Sensory evaluation of fish freshness. A reference manual for the fish industry. QIM-Eurofish (www.qim-eurofish.com).

Maturin, L. and Peeler, J.T. (2001) Chapter 3. Aerobic Plate Count. Bacteriological Analytical Manual, Gaithersgurg, MD, US Food and Drug Administration.

Mietz, J.L. and Karmas, E. (1977). Chemical quality index of canned tuna as determined by high-pressure liquid chromatography. *J. Food Sci.* 42, 155-158.

Mihaljeviæ, B., Katušin-Ra•em, B. and Ra•em, D. (1996). The reevaluation of the ferric thiocyanate assay for lipid hydroperoxides with special considerations of the mechanistic aspects of the response. *Free Radical Biology and Medicine*, 21(1): 53-63.

Miller, D.S. and Payne, P.R. (1959). A ballistic bomb calorimeter. *Br. J. Nutr.* 13, 501-508.

Nielsen, D. and Hyldig, G. (2004). Influence of handling procedures and biological factors onthe QIM evaluation of whole herring (*Clupea harengus* L.). *Food Res. Int.* 37: 975-983.

Noll, J.S., Simmonds, D.H. and Bushuk, E.C. (1974) A modified biuret reagent for the determination of protein. *Cereal Chem.*, 52, 610-616.

Nordstrom, J.L., Vickery, M.C.L., Blackstone, G.M., Murray, S.L. and DePaola, A. (2007) Development of a multiplex real time PCR assay with an internal amplification control for the detection of total and pathogenic *Vibrio parahaemolyticus* bacteria in oysters. *Appl. Environ. Microbiol.* 73: 5840-5847.

Official Methods of Analysis of AOAC International. 18th Edn. Revision 1. AOAC International, Gaithersburg, MD, 2006.

Official Methods of Analysis of AOAC International. (2006) 18th Edn. Revision 1. AOAC International, Gaithersburg, MD.

Park S, Jung J, Choi S, Oh Y, Lee J, Chae H, Ryu S, Jung H, Park G, Choi S and Kim B. (2012). Molecular characterization *of Listeria monocytogenes* based on the PFGE and RAPD in Korea. *Adv. Microbiol.* 2: 605–616.

Pearce, K.N. and Kinsella, J.E. (1978). Emulsifying properties of proteins: evaluation of a turbidimetric technique. *Journal of Agricultural and Food Chemistry*, 26(3): 716-723.

Rodrigues, T.P., Mársico, E.T., Franco, R.M., Mello, S.C.R.P., Soares, I.C., Zúniga, N.O.C. and de Freitas, M.Q. (2016). Quality index method (QIM) and quantitative descriptive analysis (QDA) of Nile tilapia (*Oreochromis niloticus*) quality indices. *African J. Agricul. Res.* 11(3): 209-216.

Rosier, J. and Petegham, C.V. (1988). A screening method for simultaneous determination of putrescine, cadaverine, histamine, spermidine and spermine in fish by means of HPLC in their 5-dimethyl amino naphthalene 1-sulphonyl derivatives. *Z. Lebensm. Unters. Forsch.* 186: 25-18,

Ryder, J.M. (1985). Determination of adenosine triphosphate and its breakdown products in fish muscle by high-performance liquid chromatography. *Journal of Agricultural and Food chemistry*, 33(4):678-680.

Saito, T., Arai, K., Matsuyoshi, M. (1959). A new method for estimating the freshness of fish. *Bull. Jpn. Soc. Sci. Fish.*, 24 (9): 749–750.

Sands, D. and Rovira, A. (1970). Isolation of fluorescence pseudomonads with a selective medium, *Appl. Microbiol.* 20: 513-514.

SANTE/11945/2015. Guidance document on analytical quality control and method validation procedures for pesticide residues and analysis in food and feed. European Commission Directorate General for Health and Food Safety. 21 – 22 November 2017.

Sato, K., Ohashi, C., Ohtsuki, K. and Kawabata, M. (1991). Type V collagen in trout (*Salmo gairdneri*) muscle and its solubility change during chilled storage of muscle. *Journal of Agricultural and Food Chemistry*, 39(7):1222-1225.

Selvaganapathi, R., Jeyasekaran, G., Jeyashakila, R., Sukumar, D., Palanikumar, M. and Sivaraman, B. (2018). Occurrence of *Listeria monocytogenes* on the seafood contact surfaces of Tuticorin coast of India. *J. Food Sci. Technol.* Doi.org/10.1007/s13197-018-3230-y.

Shewan, J.M., MacIntosh, R.G., Tucker, C.G. and Ehrenberg, A.S.C. (1953). The development of a numerical scoring system for the sensory assessment of the spoilage of wet white fish stored in ice. *Journal of the Science of Food and Agriculture*, 4(6): 283-298.

Shimamoto, J., Horatusuka, S., Hasagawa, C., Sato, M. and Kawano, S. (2003). Rapid non-destructive determination of fat content in frozen skipjack using portable near infrared spectrophotometer. *Fish Sci.* 69(4): 856-860.

Sivakumar, P., Arichandran, R., Suguna, L., Mariappan, M., Chandrakasan, G. (2000). The composition and characteristics of skin and muscle collagens from a freshwater catfish grown in biologically treated tannery effluent water. *J. Fish. Biol.* 56: 999–1012.

Solberg, C. (1997). NIR-A rapid method for quality control. In: *Seafood from Producer to Consumer, Integrated Approach to Quality*. Eds. Luten, JB, Borresen T and Ochlenshclager J. Elsevier Science, Amsterdam, the Netherlands. pp. 529-534.

Tada, J., Ohashi, T., Nishimura, N., Shirasaki, Y., Ozaki, H., Fukushima, S., Takano, J., Nishibsuchi, M. and Takeda, Y. (1992). Detection of the thermostable direct hemolysin gene (tdh) and the thermostable direct hemolysin related hemolysin gene (trh) of *Vibrio parahaemolyticus* by polymerase chain reaction. Mol. *Cell. Probes.* 6: 477-487.

Tallent, S., Hait, J., Bennet, R.W. and Gayle, A. (2016). Chapter 12. *Staphylococcus aureus*. In: Bacteriological Analytical Manual, Gaithersgurg, MD, US Food and Drug Administration.

Thirumalairaj, K., Jeyasekaran, G., Jeya Shakila, R., Thangarani, J.A., and Sukumar, D. (2011). Multiplex polymerase chain reaction assay for the detection of *Salmonella enterica* serovars in shrimps in 4 h. *J. Bacteriol. Res.* 3:56-62.

Veciana-Nogues MT, Hernandez-Jover T, Marine-FontA. (1995). Liquid chromatographic method for determination of biogenic amines in fish and fish products. *J. AOAC. Int.* 78: 1653-1655

Veliuylin, E., Van der Zaag, C., Burk, W. and Erikson, U. (2005). In vitro determination of fat content in Atlantic salmon (*Salmo salar*) with a mobile NMR spectrometer. *J. Sci. Food*. Arg. 85: 1299-1304.

Venkateswaran, K., Dohomoto, N. and Harayama, S. (1998). Cloning and nucleotide sequence of the gyrB gene of *Vibrio parahaemolyticus* and its application in detection of this pathogen in shrimp. *Appl. Environ. Microbiol.* 64: 681-687

Venkitarnarayanan, K., Khan, M., Faustman, C. and Berry, B. (1996). Detection of meat spoilage bacteria using the polymerase chain reaction. *J. Food Protect.* 59: 845-848.

Vickery, M.C., Nilsson, W.N., Strom, M.A., Nordstrom, J.L. and DePaola, A. (2007). A real time PCR assay for the rapid determination of 16S rRNA genotype in *Vibrio vulnificus.J. Microbiol. Methods.* 68: 376-384.

Virot, M., Toman, V., Colnagui, G., Visinoni, F. and Chemat, F. (2007). New microwave integrated Soxhlet extraction. An advantageous tool for the extraction of lipids from food product. *J. Chrom.* 1174, 138-144.

Vogt, A., Gormley, T.R., Downey, G. and Somers, J. (2002). A comparison of selected rapid methods for fat measurement in fresh herring (*Clupea harengus*). *J. Food. Compos. Anal.* 15, 205-251.